Print Reading
For Welding and Fabrication

Kevin Corgan
Southwestern Illinois College

Prentice Hall

Boston   Columbus   Indianapolis   New York   San Francisco   Upper Saddle River
Amsterdam   Cape Town   Dubai   London   Madrid   Milan   Munich   Paris   Montreal   Toronto
Delhi   Mexico City   Sao Paulo   Sydney   Hong Kong   Seoul   Singapore   Taipei   Tokyo

Vice President and Editor in Chief: Vernon R. Anthony
Acquisitions Editor: David Ploskonka
Editorial Assistant: Nancy Kesterson
Director of Marketing: David Gesell
Executive Marketing Manager: Derril Trakalo
Senior Marketing Coordinator: Alicia Wozniak
Marketing Assistant: Les Roberts
Project Manager: Maren L. Miller
Senior Managing Editor: JoEllen Gohr
Associate Managing Editor: Alexandrina Benedicto Wolf
Senior Operations Supervisor: Pat Tonneman

Operations Specialist: Laura Weaver
Art Director: Jayne Conte
Cover Designer: Axell Designs
Cover Image: Age Fotostock / Superstock
AV Project Manager: Janet Portisch
Full-Service Project Management: Karpagam Jagadeesan, GGS Higher Education Resources, PMG
Composition: GGS Higher Education Resources, A Division of PreMedia Global, Inc.
Printer/Binder: Quebecor Printing/Dubuque
Cover Printer: Lehigh-Phoenix Color/Hagerstown
Text Font: Palatino

Credits and acknowledgments borrowed from other sources and reproduced, with permission, in this textbook appear on the appropriate page within the text. Unless otherwise stated, all figures and tables have been provided by the author.

Library of Congress Cataloging-in-Publication Data

Corgan, K. (Kevin)
  Print reading : for welding and fabrication / K. Corgan. — 1e.
    p. cm.
  Includes bibliographical references and index.
  ISBN-13: 978-0-13-502817-9 (alk. paper)
  ISBN-10: 0-13-502817-5 (alk. paper)
  1. Blueprints. 2. Welding joints—Drawings. 3. Welding—Drawings. 4. Machinery—Drawings. I. Title.
T379.C66 2011
604.2′5—dc22

                                                                2009044133

10 9 8 7 6 5 4 3 2 1

Prentice Hall
is an imprint of

www.pearsonhighered.com

ISBN 10:   0-13-502817-5
ISBN 13: 978-0-13-502817-9

CONTENTS

PREFACE

One of the most important skills welders and fabricators can have is being able to build something correctly from reading the design requirements shown on a drawing. My goal in writing this book is to provide an easy-to-understand and logical path for students to learn to read and understand drawings that are typically found in the welding and fabrication industry. This book is intended for use at high schools, vocational schools, trade schools, and colleges, as well as by those doing independent study.

There are thirteen chapters that contain objectives, key terms, end-of-chapter exercises, and a mathematics supplement. A fourteenth chapter consists of a number of comprehensive review exercises. The mathematics supplement at the end of each chapter breaks the math skills necessary for interpreting prints into smaller, easy to digest segments of information, which also allows each institution to determine and use only the mathematics supplements that are necessary for its individual program. Emphasis is provided on developing students' awareness of the codes, standards, and other industrial practices they may encounter throughout their career. Concepts and terminology from the American Welding Society (AWS) and the Society of Mechanical Engineers (ASME) are used throughout the text.

Numerous appendixes provide acronyms and abbreviations, conversion tables, a selected pipe size chart, an ASME Y14.5 to International Standards Organizations (ISO) symbol comparison chart, an AWS master chart of welding and allied processes, and the American Welding Society welding symbol chart. A glossary defines key terms used throughout the text.

In addition to the text, there is a Companion Website (www.pearsonhighered. com/corgan), which features some of the prints from the text for further study. For the instructor, there is an Instructor's Manual and PowerPoint slides. To access supplementary materials online, instructors need to request an instructor access code. Go to **www.pearsonhighered.com/irc**, where you can register for an instructor access code. Within 48 hours after registering, you will receive a confirming e-mail, including an instructor access code. Once you have received your code, go to the site and log on for full instructions on downloading the materials you wish to use.

I would like to thank the following reviewers of this text for their time and insight: Steven K. Davis, Mt. Hood Community College; Warren Donworth, Ph.D., Austin Community College; Charlie Herberg, South Seattle Community College; Edward L. Roadarmel, Pennsylvania College of Technology; Richard J. Rowe, Johnson County Community College; Kevin Schaaf, Pandjiris, Inc.; and Larry Schaaf, Southwestern Illinois College.

It is my hope that this book will provide instructors with a logical format for their print reading course, and that it will provide students with an easy-to-use and easy-to-understand guide to reading prints and achieving their goals.

Kevin Corgan
Southwestern Illinois College

Introduction to Print Reading

OBJECTIVES

- Understand and explain who uses drawings and why
- Understand the use of the terms drawings, blueprints, and prints
- Be able to identify the parts of a drawing
- Locate areas of a drawing by using drawing zones
- Be able to identify different types of engineering drawings

KEY TERMS

Print	Drawing sheet	Mono-detail drawing
Blueprint	Title block	Multi-detail drawing
Blueline print	Bill of material	Assembly drawing
Whiteprint	Balloon	Inseparable assembly drawing
Computer-aided drafting (CAD)	Zoning coordinate system	Top drawing
	Detail drawing	Main assembly drawing

OVERVIEW

Drawings are made by engineers, designers, drafters, and others to communicate how things are to be fabricated, manufactured, or constructed. Some of the drawings are very specific and show a lot of detail, while others show very little detail. A drawing, or a copy of a drawing called a **print**, may be used by a number of different occupations by the time the design is fabricated or manufactured. For example, engineers, designers, and company owners review the designs made on drawings and copied onto prints to ensure the design meets the requirements of the project. Estimators and purchasing agents review prints for costing purposes. Weld shop and construction supervisors, lead workers, welders, machinists, carpenters, and electricians all read and use prints to understand how to properly manufacture or construct the part of the design they are required to complete.

HISTORY

Historically, engineering drawings were made by hand and then blueprint or blueline copies were made and distributed to those who needed them. The term **blueprint** comes from a type of engineering drawing copy where the background is blue and the design is shown in white. A **blueline print**, also called a **whiteprint**, is a copy of a drawing where the background is white and the design is blue. Both blueprints and blueline prints were widely used in the past but have since been replaced in most applications by modern printing methods. Today, the term *blueprint* is typically used to refer to a design or a copy of a design. With the invention of computers and the ability to develop programs for specific purposes, engineers developed software programs specifically for drawing engineering and architectural designs. This type of software is called **computer-aided drafting (CAD)**. In today's engineering field, CAD systems have become the standard method for making drawings. The drawing designs can then be printed out by using a standard printer, a large-format printer, or a plotter attached to the computer. The prints that are made from any of these methods are typically white with black lines, just like any other standard copy or printout from a copy machine or computer printer. Figure 1-1 shows a typical blackline print generated from a CAD system.

ELEMENTS OF A DRAWING

Engineering drawings are made on a **drawing sheet** that includes an outline around the perimeter of the sheet, a **title block**, and (when required) a **bill of material** area. Figure 1-2 shows a drawing sheet that includes an outline and a basic title block. Figure 1-3 (shown on page 5) is a drawing sheet that includes a more detailed title block, revision block, and bill of material. A numbered **balloon** with an arrow shown in blue points to each specific item of the drawing that corresponds to a description shown on pages 2 and 6. The information listed and the arrangement of the information listed within title blocks, bills of material, and revision blocks can vary from company to company and from project to project.

The numbered items shown in Figure 1-3 are described in detail below.

1. *Drawing perimeter:* Sets the outline for the drawing.
2. *Company name:* The name of the company who owns the drawing is placed here.
3. *Title:* This is the title (name) of the drawing.
4. *Drawing number:* Each drawing should have its own unique number. The particular numbering system used will be unique to each individual company.
5. *Revision number (current drawing):* This is where the revision number that corresponds to the current drawing is located. When a drawing is revised, it is given a revision number so that the revision can be tracked properly. When reading a drawing, make sure that you are reading the version that applies to your requirements.
6. *Sheet number:* Some drawing numbering systems have a drawing number or project number that is used for all of the drawings within the design set. They will then have different sheet numbers that go along with the project number for each drawing sheet.

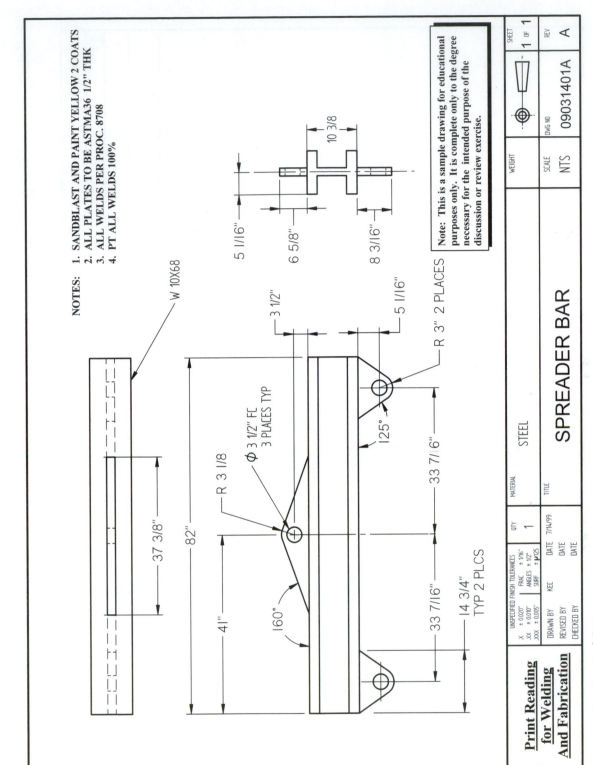

NOTES:
1. SANDBLAST AND PAINT YELLOW 2 COATS
2. ALL PLATES TO BE ASTMA36 1/2" THK
3. ALL WELDS PER PROC. 8708
4. PT ALL WELDS 100%

W 10X68

10 3/8

5 1/16"

6 5/8"

8 3/16"

3 1/2"

5 1/16"

R 3" 2 PLACES

37 3/8"

82"

Ø 3 1/2" FC
3 PLACES TYP

R 3 1/8

125°

33 7/16"

41"

160°

33 7/16"

14 3/4"
TYP 2 PLCS

Note: This is a sample drawing for educational purposes only. It is complete only to the degree necessary for the intended purpose of the discussion or review exercise.

UNSPECIFIED FINISH TOLERANCES		
X ± 0.020"	FRAC ± 1/16"	
.XX ± 0.010"	ANGLES ± 1/2°	
.XXX ± 0.005"	SURF ▽125	
DRAWN BY	KEC	DATE 7/14/99
REVISED BY		DATE
CHECKED BY		DATE

QTY			
1			

MATERIAL STEEL

TITLE

SPREADER BAR

WEIGHT		SHEET 1 OF 1
SCALE NTS	DWG NO 09031401A	REV A

Print Reading
for Welding
And Fabrication

Figure 1-1 Typical Print

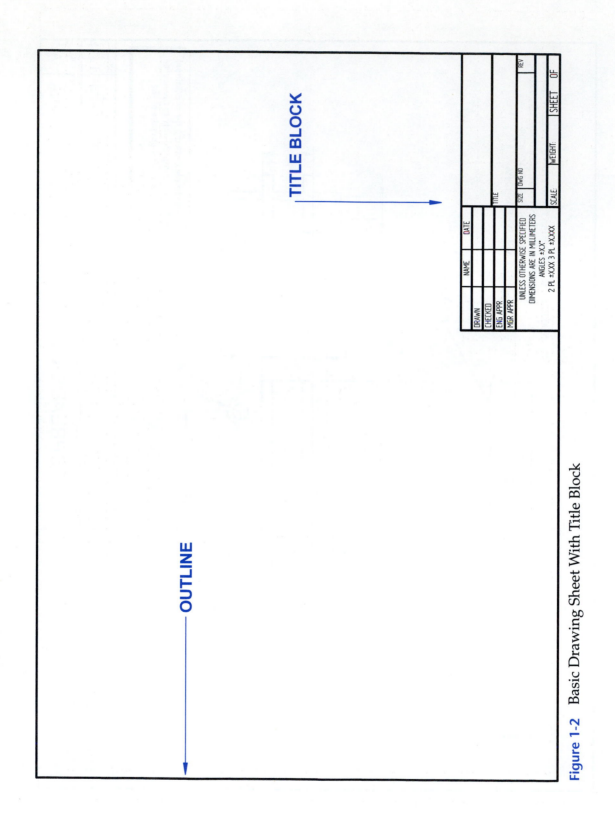

Figure 1-2 Basic Drawing Sheet With Title Block

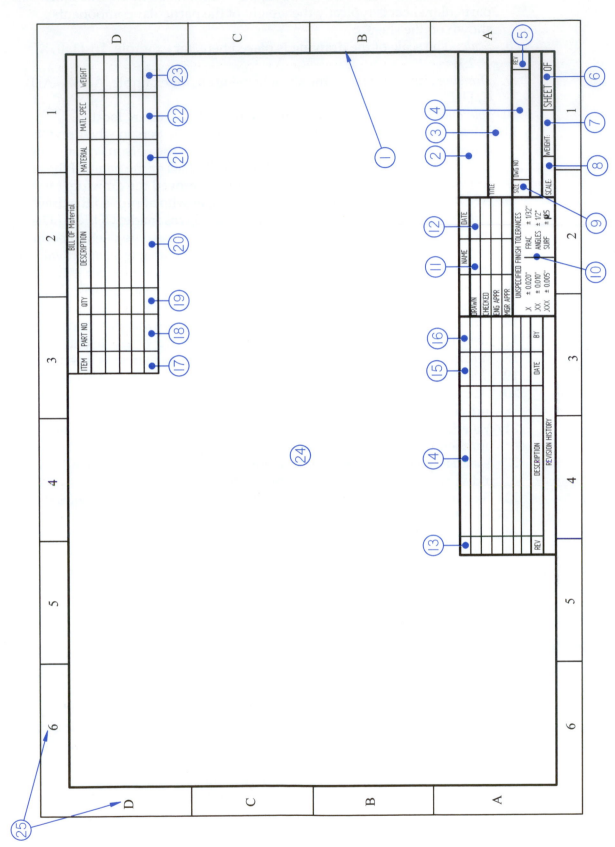

Figure 1-3 Drawing Sheet With Coordinates

7. *Weight:* Unless otherwise specified, the weight listed in the title block of a particular sheet is typically the weight of the particular components shown on the sheet.
8. *Scale:* The scale of the drawing is placed here, for example, 1:1 Half, ¼"= 1" 2:1.
9. *Drawing size:* The size of the sheet being used, for example, ANSI—A, B, C, D, E.
10. *Tolerance block:* Where general unspecified tolerances are located.
11. *Name column:* DRAWN= the initials or name of the person who drew the drawing, CHECKED = the initials or the name of the person who reviewed and checked the drawing, ENG APPR = the initials or the name of the person within the engineering department that approved the drawing, MGR APPR = the initials or the name of the manager who approved the drawing.
12. *Date column:* DRAWN = the date the drawing was made, CHECKED = the date the drawing was checked, ENG APPR= the date of engineering department approval, MGR APPR = the date of managerial approval.
13. *Revision number (history):* Column where each revision number is listed.
14. *Revision description (history):* Column where each revision is described.
15. *Revision date (history):* Column where the date of each revision is listed.
16. *Revised by:* Column showing the initials of the person who made the revision.
17. *Item:* Column where each identified item on the drawing is listed.
18. *Part number:* Column where the part number for each item is listed.
19. *Quantity:* Where the quantity corresponding to each item number is listed.
20. *Description:* The size description of the material that corresponds to the individual item number, for example, ¼"× 4" × 37".
21. *Material:* The type of material that corresponds to the individual item number, for example, CRS plate.
22. *Material specification:* Specification for the material that corresponds to the individual item number, for example, ASTM A36.
23. *Weight:* The weight of the material that corresponds to the individual item number.
24. *Drawing area:* The area where the views and notes are placed.
25. *Zoning:* A **zoning coordinate system** placed outside the outline of the sheet. If the coordinate system is provided, its purpose is to help locate specific areas or zones of the drawing, similar to the way numbers and letters are used on a map to determine location.

TYPES OF ENGINEERING DRAWINGS

As described in ASME Y14.24, Types and Applications of Engineering Drawings, there are many different types of engineering drawings that can be made to define project requirements. The most common types that are used in welding and fabrication are detail, assembly, and inseparable assembly. A **detail drawing** shows all of the information needed to completely make the part or parts shown on that particular drawing. When a single part is shown on a detail drawing, it is known as a **mono-detail drawing** (see Figure 1-4). When multiple single pieces are detailed on a single drawing, it is known as a **multi-detail drawing** (see Figure 1-5). An **assembly drawing** (shown in Figure 1-6) shows how two or more parts, other assemblies, or a combination thereof are assembled together. Weldment drawings typically fall under the category of **inseparable assembly** because most of the time, they are to be permanently joined together. See Figure 1-7.

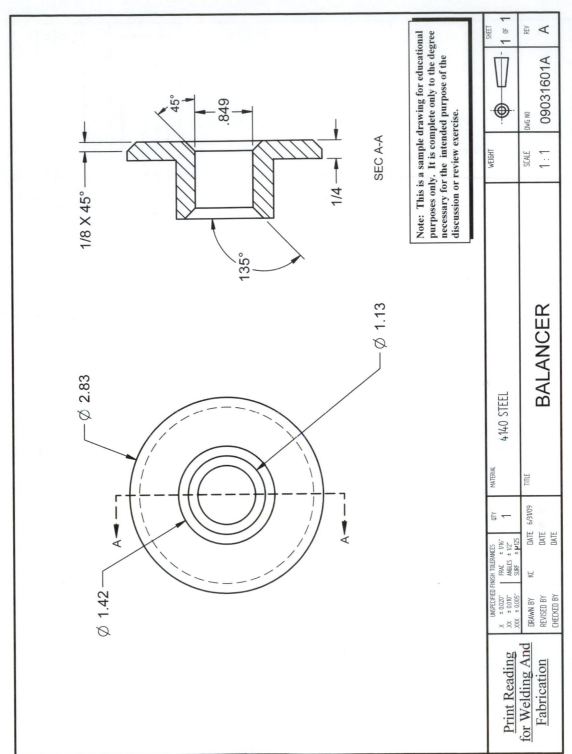

SEC A-A

1/8 X 45°
45°
.849
135°
1/4

Ø 2.83
Ø 1.42
Ø 1.13

Note: This is a sample drawing for educational purposes only. It is complete only to the degree necessary for the intended purpose of the discussion or review exercise.

UNSPECIFIED FINISH TOLERANCES				MATERIAL		
X ± 0.020"	FRAC ± 1/16"			4140 STEEL		
XX ± 0.010"	ANGLES ± 1/2°					
XXX ± 0.005"	SURF ± 125µ					
DRAWN BY	KC	DATE 6/31/09	QTY 1	TITLE		
REVISED BY		DATE				
CHECKED BY		DATE			**BALANCER**	

WEIGHT		SHEET 1 OF 1
SCALE 1:1	DWG NO 09031601A	REV A

Print Reading
for Welding And
Fabrication

Figure 1-4 Mono-detail Drawing

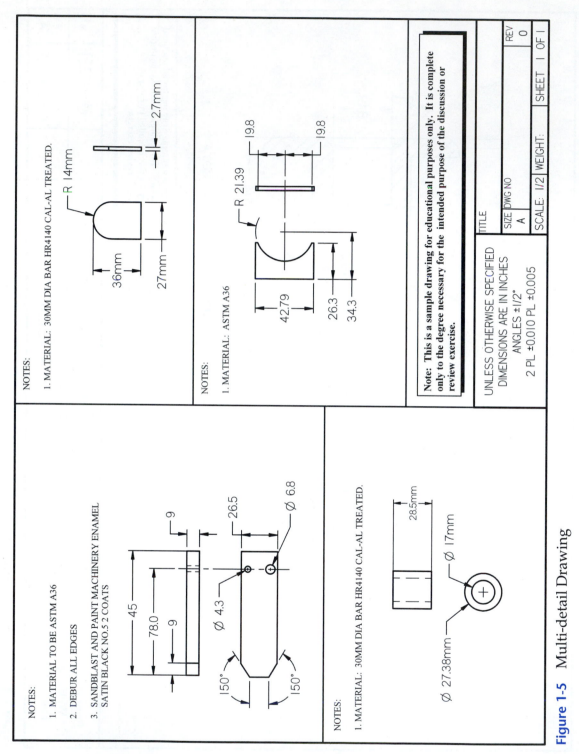

Figure 1-5 Multi-detail Drawing

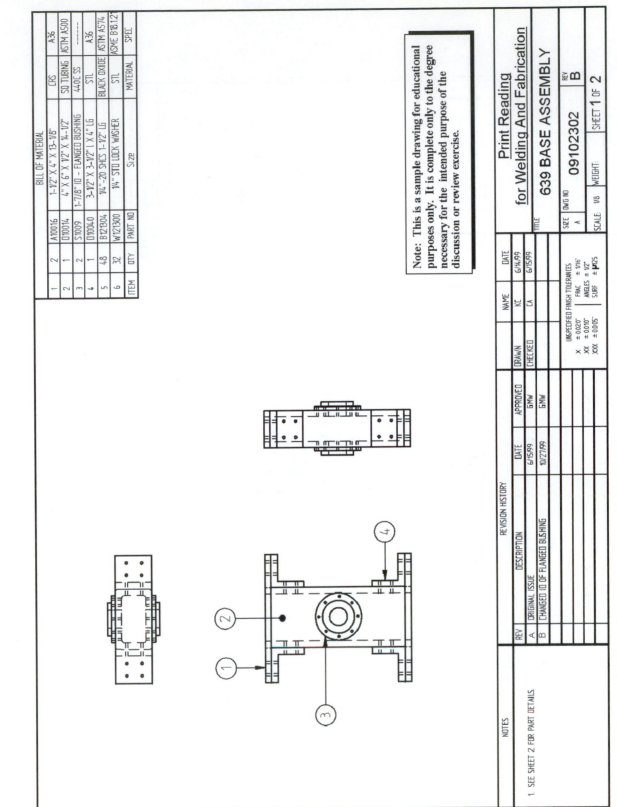

BILL OF MATERIAL

ITEM	QTY	PART NO	Size	MATERIAL	SPEC
1	2	A10016	1-1/2" X 4" X 13-1/8"	CRS	A36
2	1	D10014	4" X 6" X 1/2" X 14-1/2"	SQ TUBING	ASTM A500
3	2	S1009	1-7/8" ID – FLANGED BUSHING	440C SS	-----
4	1	D10040	3-1/2" X 3-1/2" I X 4" LG	STL	A36
5	48	B121304	1/4"-20 SHCS 1-1/2" LG	BLACK OXIDE	ASTM A574
6	32	W121300	1/4" STD LOCK WASHER	STL	ASME B18.1.2

Note: This is a sample drawing for educational purposes only. It is complete only to the degree necessary for the intended purpose of the discussion or review exercise.

Print Reading for Welding And Fabrication

TITLE 639 BASE ASSEMBLY

SIZE A	DWG NO 09102302	REV B
SCALE 1/8	WEIGHT:	SHEET 1 OF 2

	NAME	DATE
DRAWN	KC	6/14/99
CHECKED	CA	6/15/99

UNSPECIFIED FINISH TOLERANCES

X ± 0.020"	FRAC ± 1/16"
XX ± 0.010"	ANGLES ± 1/2°
XXX ± 0.005"	SURF 125

REVISION HISTORY

REV	DESCRIPTION	DATE	APPROVED
A	ORIGINAL ISSUE	6/15/99	GMW
B	CHANGED ID OF FLANGED BUSHING	10/27/99	GMW

NOTES
1. SEE SHEET 2 FOR PART DETAILS.

Figure 1-6   Assembly Drawing

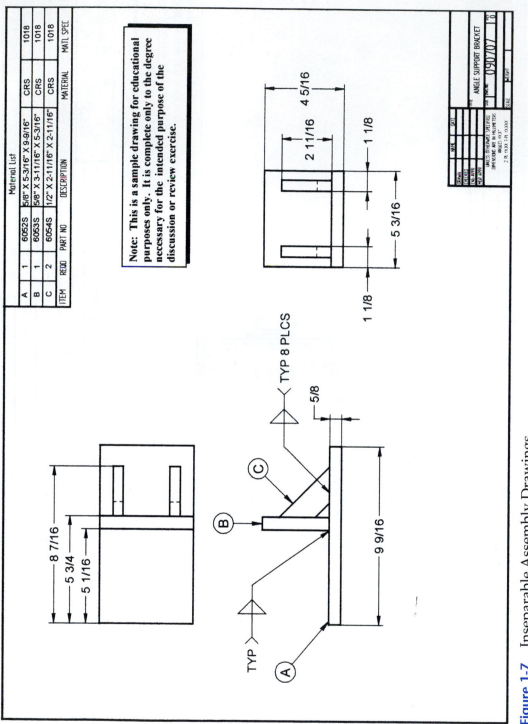

Figure 1-7 Inseparable Assembly Drawings

Some drawings include characteristics of more than one drawing type. For example, a two-piece assembly that is shown on a single drawing where both pieces are completely detailed is a combination of an assembly drawing and a detail drawing; thus, it would be considered a detail assembly drawing. When a number of individual drawings are required for a design, they are known as a drawing package or a set of drawings. Many times the first drawing of a drawing package is named the **top assembly drawing** or **main assembly drawing** because it gives an overall view of the complete assembled design.

CHAPTER 1
Practice Exercise 1

Define the following key terms.

1. Print:

2. Detail drawing:

3. Computer-aided drafting (CAD):

4. Inseparable assembly drawing:

5. Zoning coordinate system:

6. Title block:

7. Bill of material:

Practice Exercise 2

Answer the following questions.

1. Name three different occupations that use drawings.

    a. _____

    b. _____

    c. _____

2. Describe why a purchasing agent would need to review a drawing.

    _____

    _____

3. Why would a welder use a drawing?

    _____

4. How are drawings typically made today?

    _____

5. What is a main assembly drawing?

    _____

CHAPTER 1

Math Supplement

Using Addition to Find Overall Distances

During the process of reading drawings, you must often add together distances to find another distance. Example 1 shows a horizontal line broken into two segments. One segment is 53 units long; the other is 68 units long. Overall distance (A) is found by adding the two segments together.

Example 1: Find distance (A).

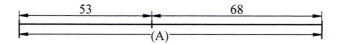

(A) = 53 + 68
(A) = 121

Example 2 requires the same basic mathematics that you used in Example 1, except that three segments are given and the addition of numbers with decimals is required.

Example 2: Find distance (A).

(A) = 28.663 + 66.963 + 25.48
(A) = 121.106

2 2 2
28.663
66.963
+ 25.480
121.106

Math Supplement Practice Exercise 1

Find distance (A) in the following.

1. (A) = _____

2. (A) = _____

3. (A) = _____

4. (A) = _____

5. (A) = _____

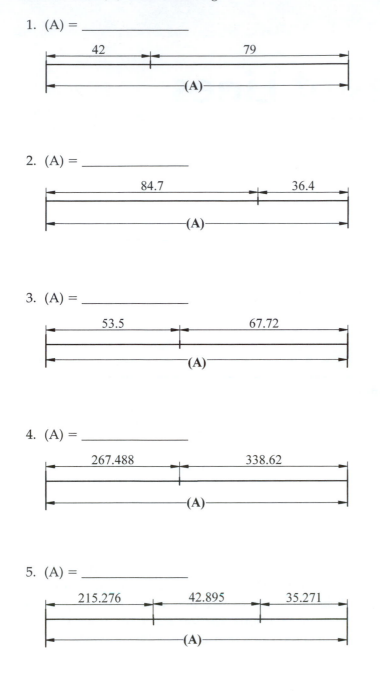

Types of Lines

OBJECTIVES

- Identify the types of lines used on drawings
- Begin to understand when, why, and how each type of line is used

KEY TERMS

Visible line	Dimension lines	Section lines
Object line	Leader	Long break lines
Hidden line	Phantom line	S-breaks
Center line	Viewing plane line	Short break line
Extension lines	Cutting plane line	Chain line

OVERVIEW

For drawings to be universally understood and to properly communicate the design to those who need it, they must follow specific guidelines. Part of the guidelines determines how lines are to be drawn and what the lines signify. Described in the following chapter sections are the standard lines used in drawings, including a description of their function and how they are to be drawn.

VISIBLE LINE

A **visible line** shows the visible edges of the object in the view (see Figure 2-1). This type of line may also be referred to as an **object line**. The lines are thick and unbroken.

Figure 2-1   Visible Line

HIDDEN LINE

A **hidden line** shows the nonvisible edges of the object from the viewpoint shown (see Figure 2-2). The lines are thin with short dashes.

- -

Figure 2-2   Hidden Line

CENTERLINE

A **centerline** is used to locate the center of holes, arcs, and other symmetrical objects (see Figure 2-3). The lines are thin lines with alternating short and long dashes.

— · —— · —— · —— · —— · —— · ——

Figure 2-3   Center Line

EXTENSION LINE

Extension lines extend from the object in the view to give location points for dimensioning (see Figure 2-4). They are thin continuous lines that are drawn with a small gap between the object and the line itself.

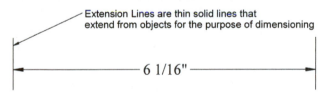

Figure 2-4   Extension Lines

DIMENSION LINE

Dimension lines are located between extension lines, center lines, other lines, or combinations thereof to show the extent of the dimension (see Figure 2-5). The measurement of the dimension is placed above, below, or in a broken area

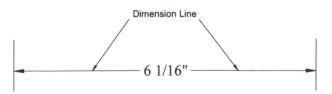

Figure 2-5   Dimension Line

of the dimension line. The ends of the dimension line are typically terminated with arrowheads. An alternate method of termination is by the use of slashes (see Figure 2-6).

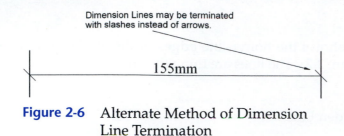

Figure 2-6   Alternate Method of Dimension Line Termination

LEADER

A **leader** is used to point to an area of the drawing were a note or dimension applies (see Figure 2-7). The leader is a fine line with either an arrow or a dot at its end. An arrow is used to point to a location. A dot is typically used to indicate a surface.

Figure 2-7   Types of Leaders

PHANTOM LINE

A **phantom line** is used to show alternate positions, mating or adjacent parts, or repeated detail (see Figure 2-8). Phantom lines are light lines made up of a series of one long and two short dashes.

———–––——–——––——–——–––—

Figure 2-8   Phantom Line

VIEWING PLANE LINES

A **viewing plane line** is used to indicate a viewpoint along a specific plane within a particular view (see Figure 2-9). The viewpoint is then shown in a separate removed view. When the plane cuts through the object, it is known as a **cutting plane line**. The separate view shown with a cutting plane line is called a section view. The lines for either viewing or cutting are thick and either broken or unbroken. Each end of the viewing or cutting plane line is angled at 90 degrees, with a short line terminating with an arrow that points in the viewpoint direction. Capital letters are placed at the ends to identify the view.

The identification letters are typically done in alphabetical order. For instance, if there is one cutting plane line on the drawing, it is identified with a capital A at the end of each arrow, signifying section A-A. If there are multiple cutting plane lines, the identification for the first is a capital A at each arrow, signifying

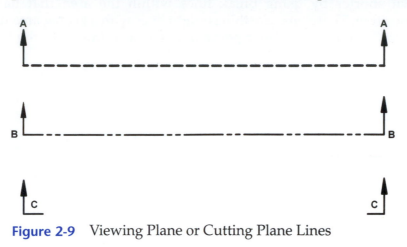

Figure 2-9 Viewing Plane or Cutting Plane Lines

section A-A; the identification of the second is a capital B at each arrow, signifying section B-B; and so on.

SECTION LINES

Section lines are lines drawn within the boundary of an object to indicate the cut area in a section view of an object. Variations in how the section lines are drawn are used to distinguish among different types of materials. Several variations are shown in Figure 2-10.

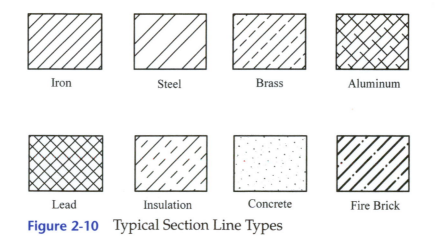

Figure 2-10 Typical Section Line Types

LONG BREAK LINE

Long break lines are used to shorten a view (see Figure 2-11). For instance, a long object that has no change in its composition over much of its length can be drawn shorter by using break lines within the area that has the same composition. These lines are thin straight lines with a zigzag area drawn in the middle. They may be either perpendicular or in line with the length of the object.

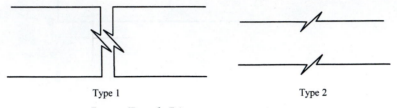

Type 1   Type 2

Figure 2-11   Long Break Line

S-BREAKS

S-breaks shorten a pipe or cylindrical tube or a cylindrical bar. The method for drawing each is shown in Figure 2-12.

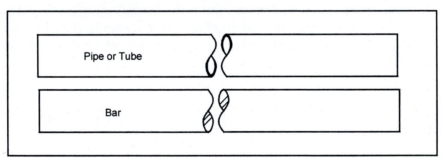

Figure 2-12   S-Break

SHORT BREAK LINE

A **short break line** is used when a short break is required, or to break out an area of a view so that another view of that specific area can be shown to clarify requirements (see Figure 2-13). The type of area that is broken out may be a partial section view, an enlarged view, or a revolved section view. Drafters sometimes use short break lines for the same purpose as long break lines.

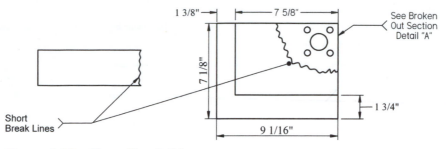

Figure 2-13   Short Break Line

CHAIN LINE

A **chain line** is used to indicate the extent and location of a surface area (see Figure 2-14). Chain lines are thick lines of long and short dashes, alternately spaced.

— · — — · — — · — — · — — · —

Figure 2-14   Chain Line

APPLICATIONS OF LINES

Figure 2-15 is a good reference because it shows line types and how they are used. It is found in ASME Standard Y14.2-2008, Line Conventions and Lettering.

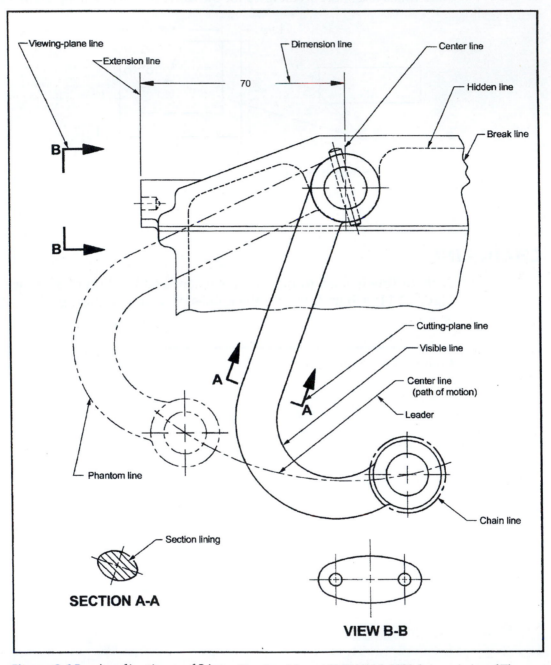

Figure 2-15   Applications of Lines *(Reprinted from ASME Y14.2-2008, by permission of The American Society of Mechanical Engineers. All rights reserved.)*

CHAPTER 2
Practice Exercise 1

Fill in the following chart by drawing the line type and briefly describing the purpose for each corresponding line.

Line Name	Line	Purpose
Visible Line		
Hidden Line		
Center Line		
Cutting Plane Line		
Steel Section Lines		
Extension Lines		
Dimension Line		
Leader Lines		
Long Break Line		
Short Break Line		
S-Break		
Phantom Line		
Chain Line		

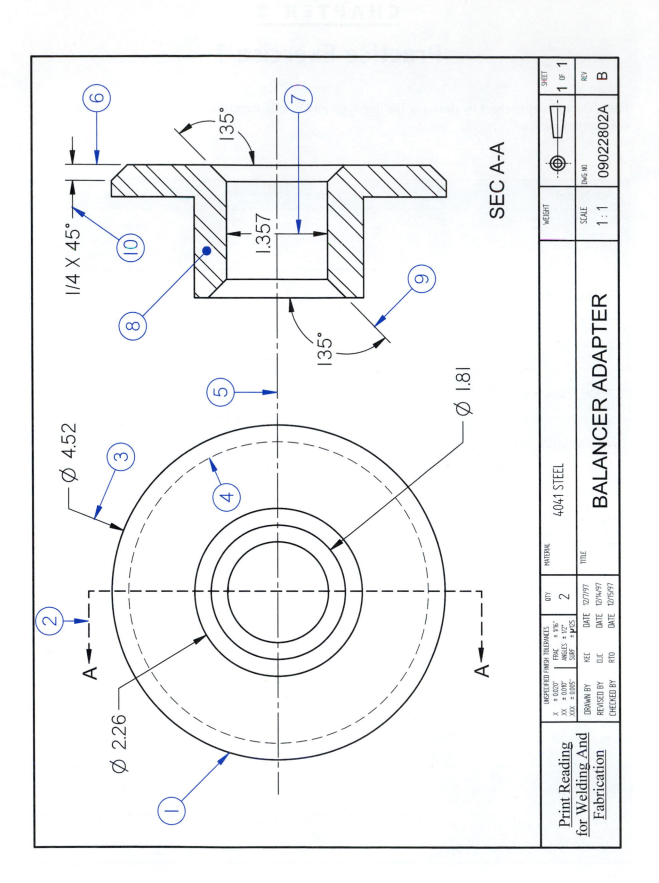

SEC A-A

135°

1.357

135°

1/4 X 45°

Ø 4.52

Ø 1.81

Ø 2.26

A

A

UNSPECIFIED FINISH TOLERANCES		QTY	MATERIAL		WEIGHT		SHEET
X ± 0.020" FRAC ± 1/16"		2	4041 STEEL				1 OF 1
XX ± 0.010" ANGLES ± 1/2"							
XXX ± 0.005" SURF ✓125		TITLE					

DRAWN BY	KEC	DATE	12/7/97
REVISED BY	DJC	DATE	12/14/97
CHECKED BY	RTO	DATE	12/15/97

BALANCER ADAPTER

SCALE 1:1   DWG NO 09022802A   REV B

Practice Exercise 2

Refer to drawing number 09022802A REV B for the following questions.
Name the lines 1 through 10 as shown on the drawing for questions 1 through 10.

1. _____

2. _____

3. _____

4. _____

5. _____

6. _____

7. _____

8. _____

9. _____

10. _____

11. What is the title of this drawing?

_____

12. What kind of material is the part made from?

_____

13. Who made this drawing?

_____

14. Who checked this drawing?

_____

15. What revision is this drawing?

_____

CHAPTER 2
Math Supplement

Using Subtraction to Find a Distance

Many times during the process of reading drawings, the difference between distances must be found. Example 1 shows a horizontal line broken into two segments. One segment is 55 units long; the other is unknown. The overall distance of 121 units is given. Distance (A) is found by subtracting the known distance of 55 units from the overall distance of 121.

Example 1: Find distance (A).

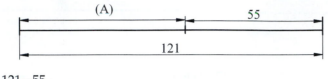

(A) = 121 − 55
(A) = 66

Example 2 requires the same basic mathematics as Example 1 except that three segments are given and the subtraction of numbers with decimals is required. An easy way to find the answer is to add the two known segments together and then subtract that total from the overall length.

Example 2: Find distance (A).

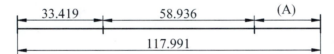

Step 1: 33.419 + 58.936 = 92.335

Step 2: 117.991 − 92.335 = 25.656

(A) = 25.656

Math Supplement Practice Exercise 1

Find distance (A) in each of the following.

1. (A) = _____

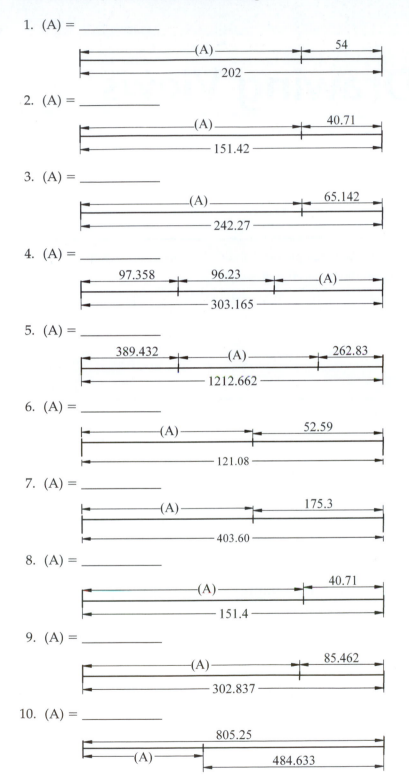

2. (A) = _____

3. (A) = _____

4. (A) = _____

5. (A) = _____

6. (A) = _____

7. (A) = _____

8. (A) = _____

9. (A) = _____

10. (A) = _____

Basic Drawing Views

OBJECTIVES

- Identify and understand the difference between pictorial and orthographic projection
- Identify the basic views in an orthographic projection drawing
- Demonstrate the ability to locate and determine the lines required on drawings

KEY TERMS

Orthographic projection	Oblique	Plan view
Pictorial	Six principal views	Elevation view
Isometric		

OVERVIEW

Engineering drawings for weldments can be identified by two main categories: **orthographic projection** and **pictorial**. Pictorial drawings are further broken down into two subtypes: **isometric** and **oblique**. Figure 3-1 shows each of these types of drawings.

ORTHOGRAPHIC PROJECTION

Orthographic projection drawings are the best types of drawing for showing the true size, shape, and features of objects. They are the most common drawings used in welding and fabrication. Figure 3-2 shows the **six principal views** (front, top, right side, left side, bottom, and back) that may be used. As you can see, each view shows each side of the object from a straight-on

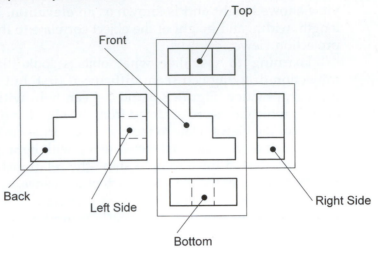

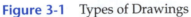

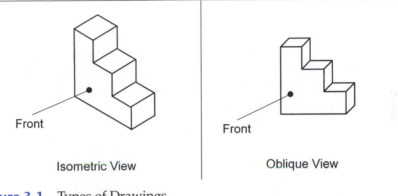

Figure 3-1 Types of Drawings

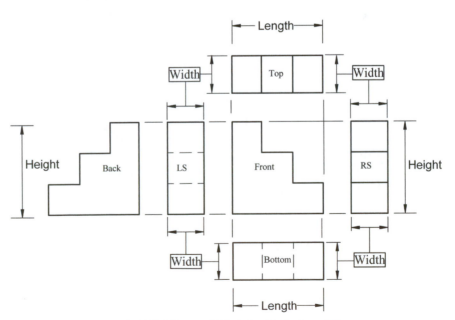

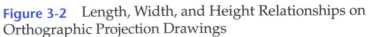

Figure 3-2 Length, Width, and Height Relationships on Orthographic Projection Drawings

viewpoint. The top view is also known as the **plan view**; a front, back or side view shows height and is known as an **elevation**. Figure 3-2 shows how the length, width, and height of the object correlate to the 6 principal orthographic projection views.

Learning to visualize what objects look like from an orthographic projection drawing can seem difficult at first, but this skill can be achieved through practice. Figure 3-3 and 3-4 can help with understanding how the orthographic views are obtained from the object and how the lines and corners in each of the views relate to the lines and corners in the other views. Figure 3-4 shows each of the principal views, along with light gray lines with arrows and light gray dotted lines to help show the relationship between the lines and points (features of the object) in each of the views.

When an engineer or a drafter is making a drawing, he or she determines which views are required to properly communicate the requirements to those who will read it. Although an orthographic projection drawing can contain views of all sides (front, top, right side, left side, bottom, and back) of an object, the least amount of views required to thoroughly communicate the requirements is typically used. (*Note:* The front, top, and right-side views are the most common views used.) The view that best illustrates the shape of the object is typically chosen as the first view that is drawn. This view is normally identified as the front view regardless of whether or not it is actually the front of the object. Once the front view has been chosen, other views, if required, are chosen based on trying to use the views in which the least amount of hidden lines will be required. As you can see in Figure 3-5, the left-side, bottom, and

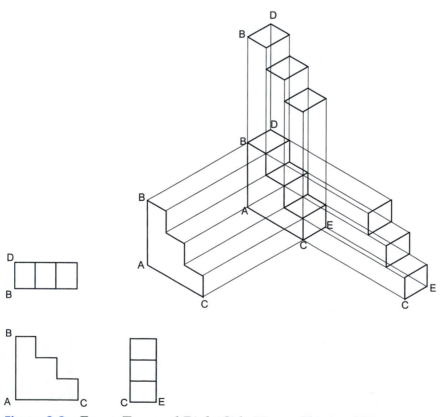

Figure 3-3 Front, Top, and Right Side Views Obtained From an Isometric View

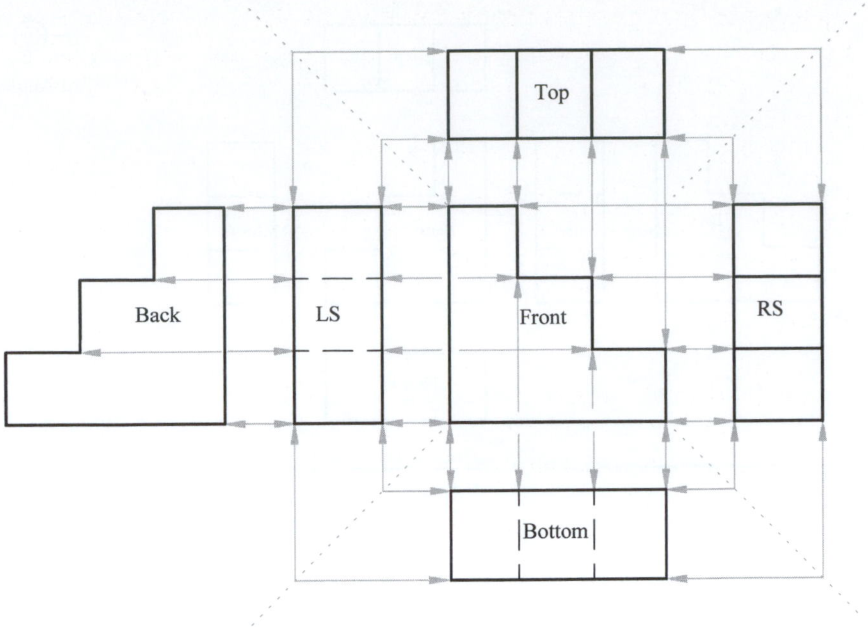

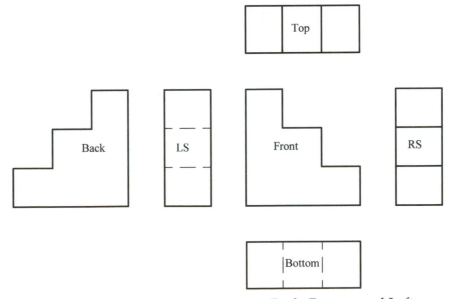

Figure 3-4   Common Points in Orthographic Projection Drawings

back views are not necessary for this drawing in order to provide the reader with the required information. For this reason, they can and should be left off.

Thus far, the arrangement of the views in the orthographic projection drawings shown in each of the figures has been what is known as third-angle projection. In simple terms, this means that the standard arrangement of views is that the right side is to the right of the front view, the left side is to the left of the front view, the top is above the front view, the bottom is below the front view, and the back is to the left of the left-side view. Although variations and alternate placement of views can occur, this arrangement of the views is the standard arrangement typically used in the United States. First-angle projection

Figure 3-5   Drawing with Unnecessary Back, Bottom and Left Side Views

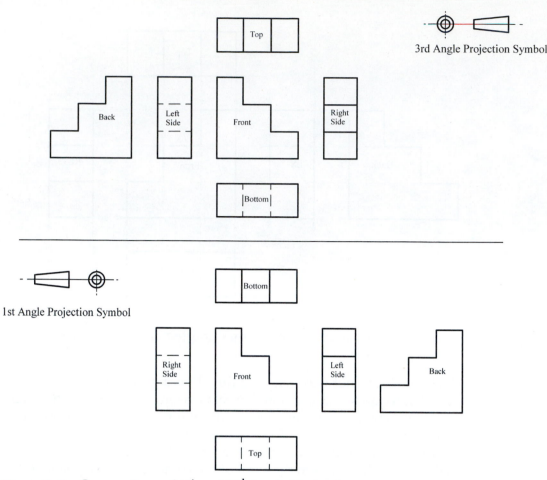

Figure 3-6   Comparison of 1ˢᵗ and 3ʳᵈ Angle Projection

is another type of orthographic projection used in many other countries. The difference between third-angle projection and first-angle projection is that the right- and left-side view locations are reversed, the top and bottom view locations are reversed, and the back view is moved to the opposite side. A comparison of third- and first-angle projections is shown in Figure 3-6.

PICTORIAL DRAWINGS

Pictorial views are often used in simple sketches and as an added view on orthographic projection drawings to help the reader better visualize the part. While they do give a three-dimensional view of an object, they are limited in their ability to show the object's true size and shape. This is demonstrated in Figure 3-7, where the circular hole that is drawn in the isometric block drawing looks circular on the block, but in fact its true shape is an ellipse. When the hole is drawn, its true shape and size on the isometric drawing doesn't look right. This is due to the fact that pictorial drawings are drawn at angles to give a three-dimensional simulation of an object.

An isometric drawing is a type of pictorial drawing that is made at 30-degree angles from a point located on a horizontal line. An oblique drawing is a type of pictorial drawing that is made with the front view in line with a

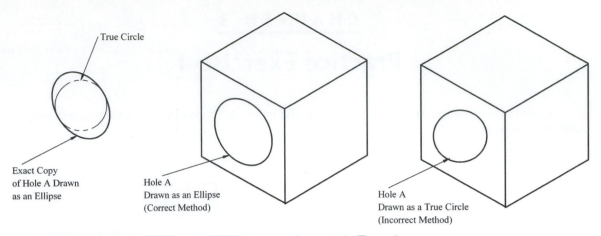

True Circle

Exact Copy
of Hole A Drawn
as an Ellipse

Hole A
Drawn as an Ellipse
(Correct Method)

Hole A
Drawn as a True Circle
(Incorrect Method)

Figure 3-7   Representation of a Hole on an Isometric Drawing

horizontal line and either the left or right side typically oriented at a 30 or 45-degree angle to the front view. See Figure 3-8. The width on an oblique view may be shown shorter than its true width. Both types of pictorial views typically do not include hidden lines.

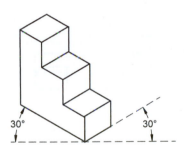

30°   30°

Isometric View

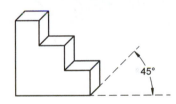

45°

Oblique View

Figure 3-8   Isometric and Oblique Views

Practice Exercise 1

Draw a line connecting the isometric drawing on the left to its corresponding orthographic projection views shown on the right.

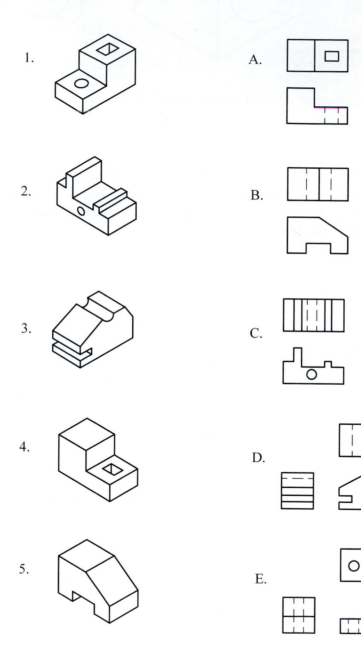

1.

2.

3.

4.

5.

A.

B.

C.

D.

E.

Practice Exercise 2

The following isometric drawings are missing lines or portions of lines. Look at the corresponding orthographic projection drawings to determine which lines or line portions are missing and draw them in.

1.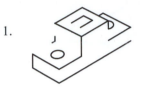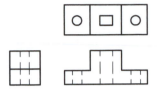

2.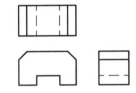

3.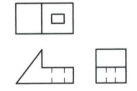

CHAPTER 3

Practice Exercise 3

Review the drawing below to answer the questions on pages 36 and 37.

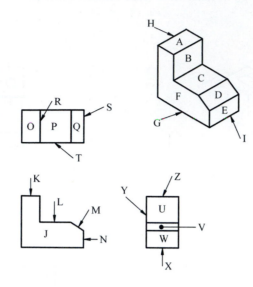

1. Surface A is represented by which surface in the top view? _____

2. Line R represents which surface shown in the right-side view? _____

3. Surface Q is represented by which line in the front view? _____

4. Line N represents which surface shown in the right-side view? _____

5. Surface P is represented by which line in the front view? _____

6. Surface U is represented by which surface in the isometric view? _____

7. Line T represents which surface in the isometric view? _____

8. What line represents surface B? _____

9. Surface E is represented by which line in the top view? _____

10. Surface E is represented by which line in the front view? _____

11. Surface D is represented by which surface in the top view? _____

12. Surface D is represented by which surface in the right-side view? _____

13. Line M represents which surface in the right-side view? _____

14. Surface J is represented by which line in the top view? _____

15. Line Y represents which surface shown in the front view? _____

16. Line I is represented by which line in the right-side view? _____

17. Surface U is represented by which line in the top view? _____

18. Surface V is represented by which surface in the top view? _____

19. Surface C is represented by which surface in the top view? _____

20. Line K represents which surface in the isometric view? _____

CHAPTER 3

Practice Exercise 4

The following orthographic drawings are missing lines or portions of lines. Look at each of the corresponding views to determine which lines or line portions are missing and draw them in.

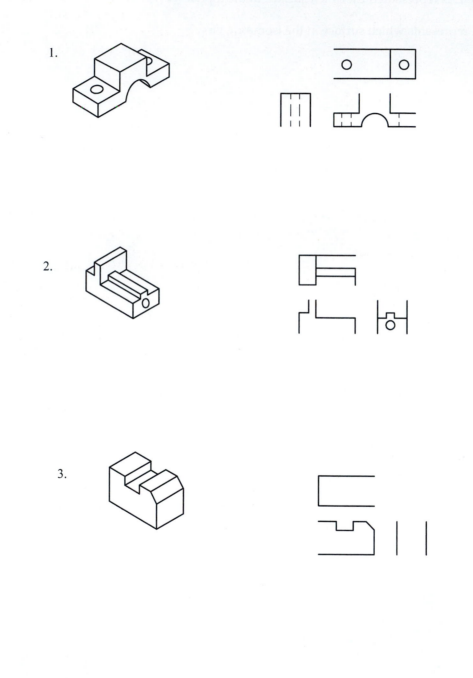

1.

2.

3.

CHAPTER 3

Math Supplement

Sixteenth and Thirty-Second Scales on Measuring Devices

Sixteenth and thirty-second scales are commonly found on measuring devices such as tape measures and steel rules used in typical fabrication activities. The knowledge of how to read these two scales can be transferred to any other common increment scale (i.e., sixty-fourth scale). Typical sixteenth and thirty-second scales are shown in Figure 3-9, on the drawing of a typical 6" steel rule.

Figure 3-9   Typical 6" Steel Rule

    A sixteenths scale means that there are sixteen (16) equal spaces marked for every inch. A thirty-second scale means that there are thirty-two (32) equal spaces marked for every inch. This is shown in Figure 3-10.

    The graduations on the measuring device have different length lines that are based on the lowest common denominator (LCD) for the fractional portion of the inch that the line represents. On a sixteenths-style scale, the sixteenth lines are the shortest, the eighth lines are the next longest, the one-quarter lines are the next longest, the one-half lines are next longest, and the full-inch marks go completely across the scale. When the measurement is taken or read, the lowest common denominator fraction (LCD FRACTIONS shown in Figure 3-11) for the graduation is used.

    On a thirty-second-style scale, the thirty-second lines are the shortest, the sixteenth lines are a little longer, the eighth lines are the next longest, the one-quarter lines are the next longest, the one-half lines are the next longest, and the full-inch marks go completely across the scale. See Figure 3-12.

    When making a measurement that is longer than 1" with a sixteenth or thirty-second scale, add the whole number to the fractional number to obtain the final value. See Figure 3-13.

    When the measurement goes over 12", the measurement can be written in total inches plus the fraction of an inch. It can also be written as feet, inches, and the inch fraction. See Figure 3-14.

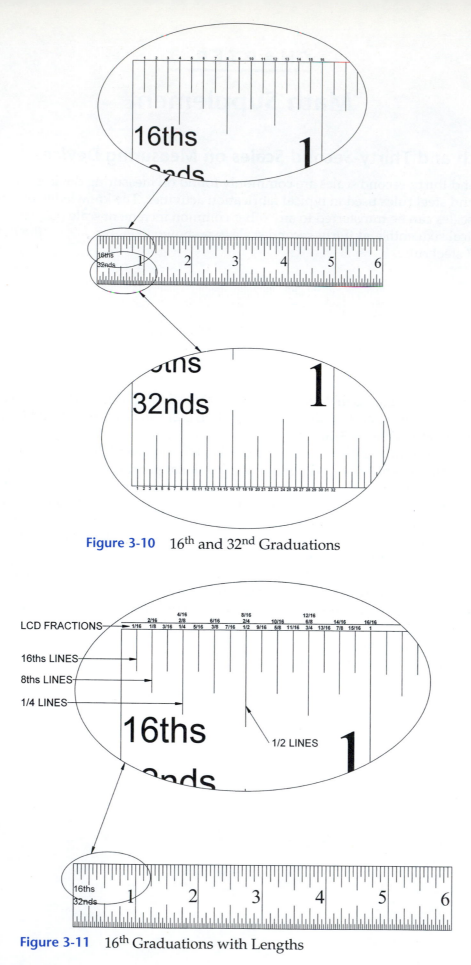

Figure 3-10 16th and 32nd Graduations

Figure 3-11 16th Graduations with Lengths

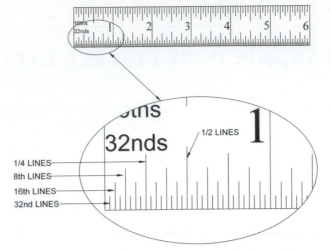

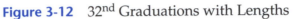

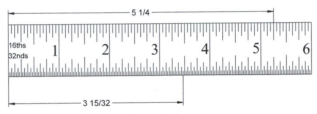

Figure 3-12   32nd Graduations with Lengths

Figure 3-13   Inch Plus Measurements

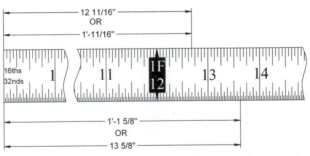

Figure 3-14   Measurements Greater Than 12″

CHAPTER 3

Math Supplement Practice Exercise 1

Determine the following lengths and write them in LCD form. For the measurements that go over the 1' mark, give the answer in both feet and inches and inches only form.

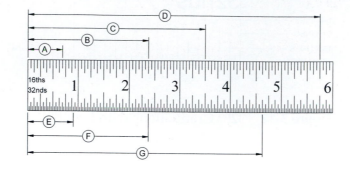

A _____

B _____

C _____

D _____

E _____

F _____

G _____

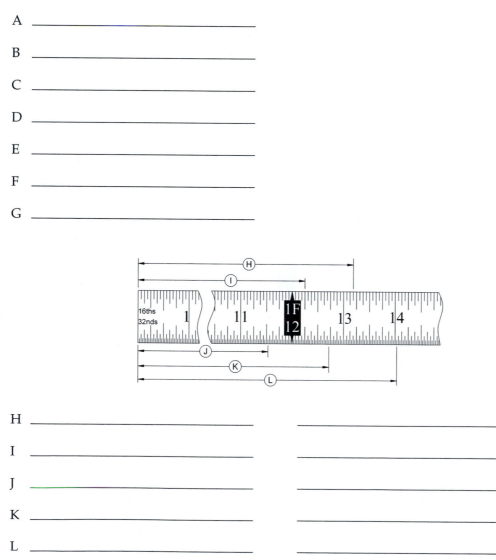

H _____   _____

I _____   _____

J _____   _____

K _____   _____

L _____   _____

Dimensions

KEY TERMS

Dimension	Limit	Tabular dimensioning
Linear dimension	Least material condition	Baseline dimensioning
Angular dimension	(LMC)	Conventional dimensioning
Reference dimension	Maximum material condition	Chain dimensioning
Tolerance	(MMC)	Broken chain dimensioning

OVERVIEW

A **dimension** gives size, location, orientation, form, or other characteristic information for the objects shown on the drawing. Figure 4-1 and 4-2 show the difference between a drawing without dimensions and one with dimensions. As you can see in Figure 4-1, the drawing gives the reader the shape of the object but doesn't give enough information to manufacture the object to its proper size. Figure 4-2 shows the same drawing including dimensions that provide the size and location information necessary to make the part.

Figure 4-1 Drawing Without Dimensions

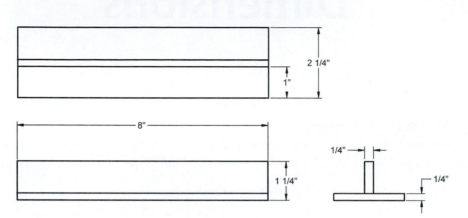

Figure 4-2 Drawing With Dimensions

LINEAR DIMENSIONS

A dimension that provides length details on objects is called a **linear dimension**. This type of dimension may be used to show either size or location and is typically given in either U.S. standard units or in metric units. Typical U.S. standard units for weld fabrication drawings will be in fractions of an inch, inches, feet, or some combination of these. When using metric units on fabrication drawings, the units are typically given in millimeters (mm), centimeters (cm), or meters (m). Figure 4-3 shows linear dimensions on an object; the linear dimensions shown give both size and location with various types of units.

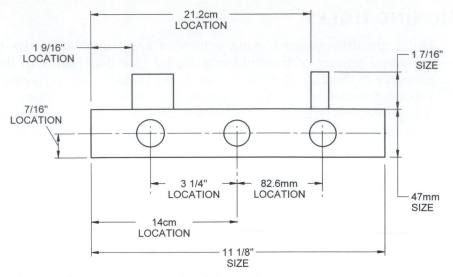

Figure 4-3   Size and Location Dimensions

ANGULAR DIMENSIONS

A dimension that provides details on the size of an angle is called an **angular dimension**. The unit for these dimensions is the degree. Figure 4-4 shows angular dimensions that are given in whole degree units. When fractions of a degree are required, they are given in either decimal fraction, common fraction, or minutes and seconds. Figure 4-5 shows examples of each.

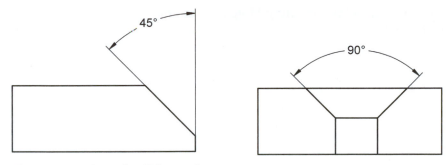

Figure 4-4   Angular Dimensions

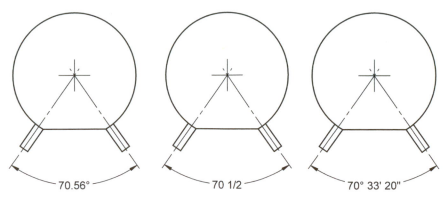

Figure 4-5   Angular Dimensions with Decimal, Common Fraction, and Minute, Second Format

DIMENSIONING HOLES

Holes are dimensioned using extension and dimension lines to show the outside diameter of the hole or a leader line that points to the hole with reference to the diameter. Figure 4-6 shows examples of each of these methods. When information specifying requirements other than the diameter of the hole is required, it is typically given following the size on a leader-type dimensioned hole. Chapter 5 of this text describes specifications for holes in greater detail.

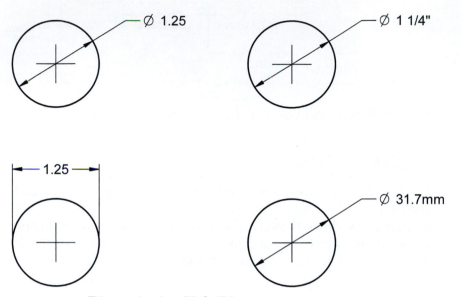

Figure 4-6   Dimensioning Hole Diameters

DIMENSIONING SLOTS

A **slot** is basically an elongated hole. It may be rectangular or have a radius at each end. Examples of each type, along with several different methods of placing dimensions are shown in Figure 4-7.

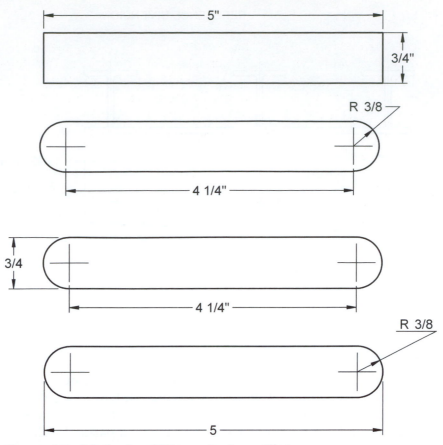

Figure 4-7   Methods of Dimensioning a Slot

REFERENCE DIMENSIONS

A dimension followed by the abbreviation REF, or one that is shown in parentheses by itself in the dimension line, is called a **reference dimension** (see Figure 4-8). This type of dimension is one that can already be found mathematically by the reader through the use of other dimensions shown on the drawing. Reference dimensions are usually placed on drawings for the convenience of the reader and are not to be used for any type of accurate measurement during layout, production, or inspection.

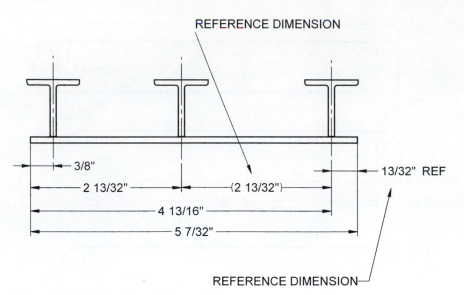

Figure 4-8   Reference Dimensions

TOLERANCES

Tolerance is the amount of variance from the dimension shown that is acceptable (see Figure 4-9). In other words, tolerances give the amount over and/or under the given dimension that the part can be made. For example, a tolerance of ±⅛" applied to a 6" dimension on a particular drawing means that the part will be acceptable if it is made anywhere between and including 5⅞" and 6⅛". The minimum or maximum that a dimension can be after adding the applicable tolerance to the dimension is known as the **limit**. The lower limit in the example above is 5⅞", and the upper limit on the example above is 6⅛".

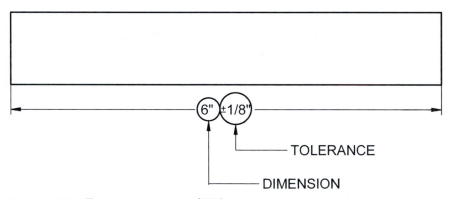

Figure 4-9   Representation of Tolerance

Every non-reference dimension on the drawing should have a tolerance that applies to it. The tolerance may be found within the dimension line, along with the dimension, as shown above in Figure 4-9, or it may be found in a general note or supplementary block on the drawing. Tolerances given in a supplementary block on the drawing are general tolerances and apply to dimensions that do not have a tolerance otherwise specified.

METHOD OF SHOWING ACCEPTABLE SIZE

Figure 4-10 shows a part with several different methods of showing the parts size requirement and the tolerances that are to be applied. The first dimension is shown without a tolerance located in the area of the dimension. In this case, look to the general tolerances listed for the drawing to find the applicable tolerance. The second dimension has a ± tolerance directly following the dimension. The third and fourth methods define the size requirements by giving the lower and upper limits of the size.

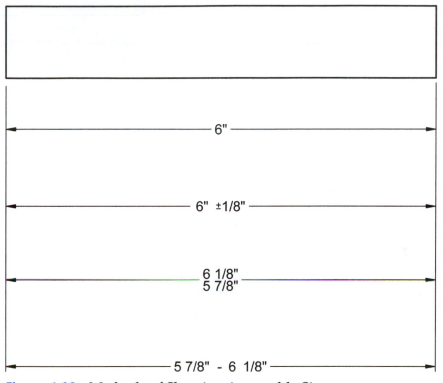

Figure 4-10   Methods of Showing Acceptable Size

LEAST AND MAXIMUM MATERIAL CONDITION

The condition where the least amount of material exists at the maximum extent of the tolerance application is called **least material condition (LMC)** (see Figures 4-11 and 4-12). In the case of an external measurement, such as the length of an object, it would occur when the minus tolerance is added to the basic dimension. In the case of a hole, it would occur when the plus tolerance is added to the basic dimension. **Maximum material condition (MMC)** is the condition where the maximum amount of material exists at the maximum extent of tolerance application (see Figures 4-11 and 4-12). In the case of an external measurement, such as the length of an object, it would occur when the plus tolerance is added to the basic dimension. In the case of a hole, it would occur when the minus tolerance is added to the basic dimension.

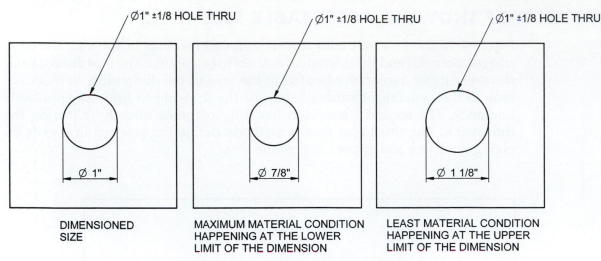

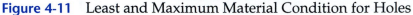

Figure 4-11 Least and Maximum Material Condition for Holes

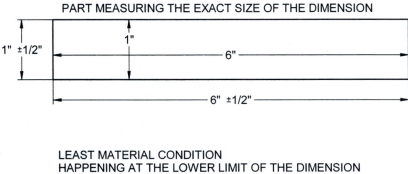

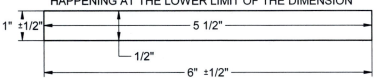

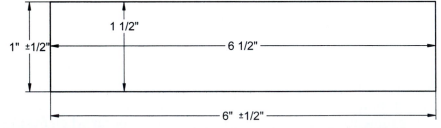

Figure 4-12 Least and Maximum Material Condition for Length and Width

TABULAR DIMENSIONING

Tabular dimensioning is a method of dimensioning used so that one drawing can be made for various parts that are similar but have differences in their size (see Figure 4-13). Letters are placed within the dimension lines on the print in

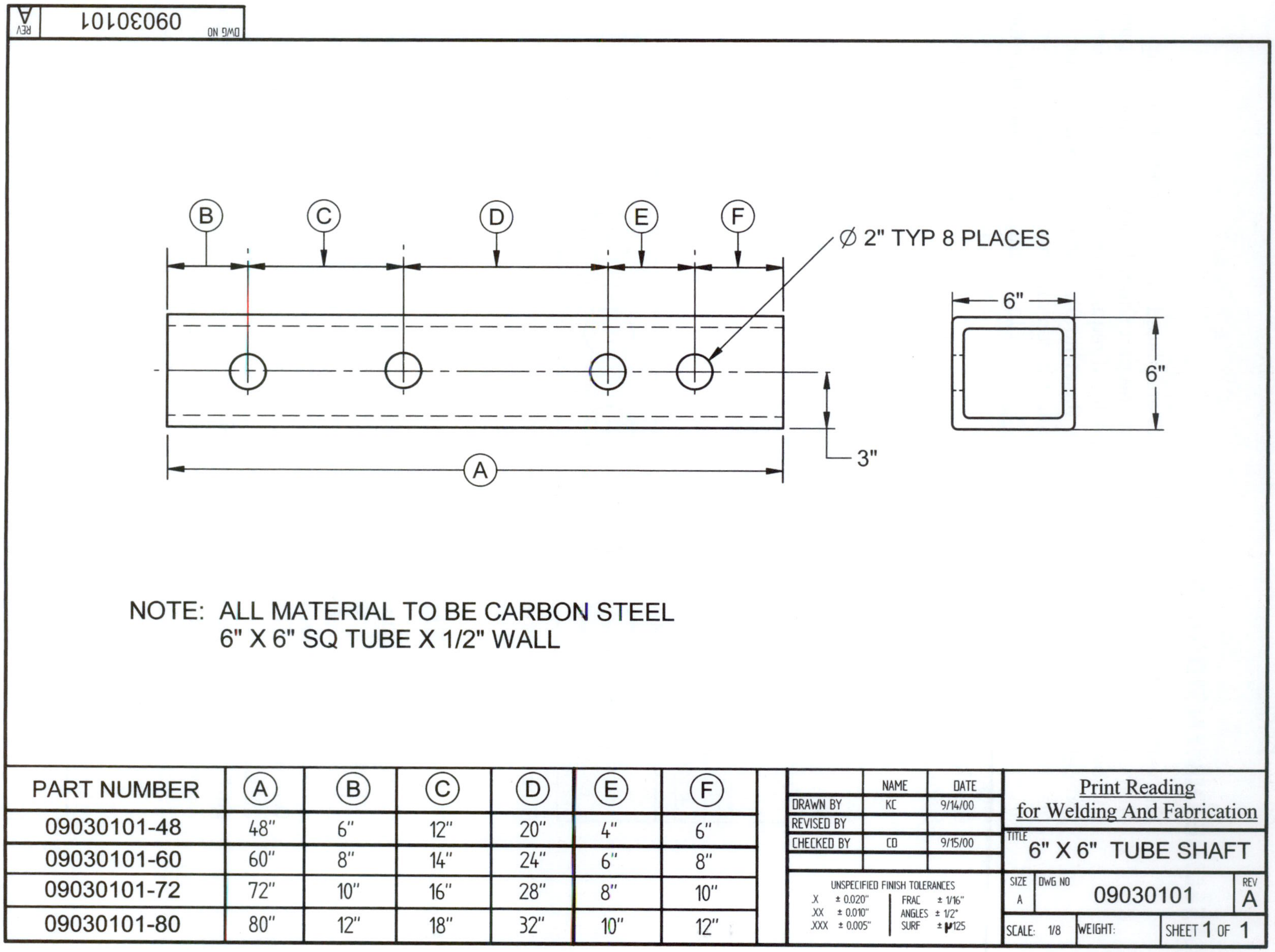

Ø 2" TYP 8 PLACES

6"

6"

6"

3"

NOTE: ALL MATERIAL TO BE CARBON STEEL
6" X 6" SQ TUBE X 1/2" WALL

PART NUMBER	A	B	C	D	E	F
09030101-48	48"	6"	12"	20"	4"	6"
09030101-60	60"	8"	14"	24"	6"	8"
09030101-72	72"	10"	16"	28"	8"	10"
09030101-80	80"	12"	18"	32"	10"	12"

	NAME	DATE
DRAWN BY	KC	9/14/00
REVISED BY		
CHECKED BY	CD	9/15/00

Print Reading
for Welding And Fabrication

TITLE 6" X 6" TUBE SHAFT

SIZE	DWG NO	REV
A	09030101	A

SCALE: 1/8   WEIGHT: SHEET 1 OF 1

UNSPECIFIED FINISH TOLERANCES
X   ± 0.020"   FRAC   ± 1/16"
XX   ± 0.010"   ANGLES ± 1/2°
XXX ± 0.005"   SURF   ± 125

DWG NO
09030101

REV
A

Figure 4-13 Tabular Dimensioning

Dimensions   51

place of actual dimensions. The actual dimensions for each different sized part are then listed below or beside their corresponding letter within a table shown on the print.

BASELINE DIMENSIONING

The way the dimensions and tolerances are placed on a drawing has an impact on the accumulation of tolerances, how the part should be made, how the part will be checked for meeting the dimensional requirements, and what the final geometry and size of the part will be. **Baseline dimensioning** is a method of placing dimensions so that dimensions in a given direction for a particular part originate from a single baseline. This is shown in Figure 4-14. Notice that the longest this piece can be is 7⅟₁₆ (found by adding one ⅟₁₆ plus tolerance to 7) and the shortest this piece can be is 6¹⁵⁄₁₆ (found by subtracting one ⅟₁₆ minus tolerance from 7).

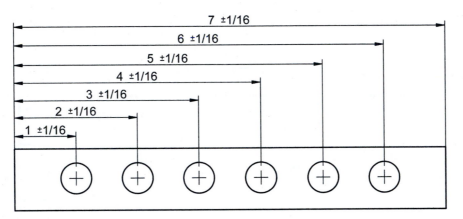

Figure 4-14 Baseline Dimensioning

CONVENTIONAL DIMENSIONING

Conventional dimensioning, also known as **chain dimensioning**, is where dimensions in any one direction for an object do not originate from a single extension line. At least one extension line from each of the dimensions in chain dimensioning originates from another dimension or dimensioned location. When using this method of dimensioning, the tolerances accumulate to allow the same part shown in Figure 4-14 to be much larger or smaller. The longest this part can be made, as dimensioned in Figure 4-15, is 7⁷⁄₁₆ (found by adding seven ⅟₁₆ plus tolerances to 7), and the shortest length this part can be made is 6⁹⁄₁₆ (found by subtracting seven ⅟₁₆ minus tolerances from seven).

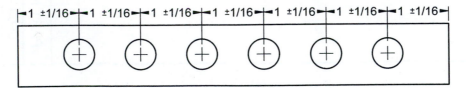

Figure 4-15 Conventional/Chain Dimensioning

BROKEN CHAIN DIMENSIONING

Another method of dimensioning, called **broken chain dimensioning**, dimensions an overall length and all but one of the individual partial lengths. The distance that is left without a dimension is the area that will accumulate all of the tolerances that build up from the other dimensions. The example shown in Figure 4-16 has an overall length that is held to a maximum of 7⅟₁₆, and a minimum overall length that is held to 6¹⁵⁄₁₆. The area left without a dimension can measure from a minimum of ⁹⁄₁₆ (obtained from subtracting the maximum chain dimension lengths of 6⁶⁄₁₆ from the minimum overall length of 6¹⁵⁄₁₆) to 1⁷⁄₁₆ (obtained from subtracting the minimum chain dimensioned added lengths of 5¹⁰⁄₁₆ from the overall maximum permissible length of 7⅟₁₆).

Note: Drawings can include any or all types of dimensioning methods.

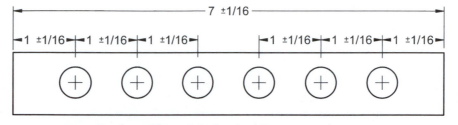

Figure 4-16   Broken Chain Method of Dimensioning

Practice Exercise 1

Define the following key terms.

1. Dimension

2. Linear dimension

3. Angular dimension

4. Reference dimension

5. Tolerance

6. Limit

7. Maximum material condition (MMC)

8. Least material condition (LMC)

9. Tabular dimensioning

10. Baseline dimensioning

11. Conventional dimensioning

12. Chain dimensioning

13. Broken chain dimensioning

Math Supplement

Fractions

Fractions are portions of whole numbers. They may be written in either common fraction form or in decimal form. Fractions such as ⅜ or ⁹⁄₁₆ are common fractions, while .375 and .5625 are decimal fractions. Common fractions have a numerator and a denominator, with a line separating them. For example, the fraction ½ consists of a numerator (1), a denominator (2), and a divisor line separating them. The fraction is typically read "one-half," but it can also be read as "one over two" or "1 divided by 2," such as in the case of converting the common fraction to a decimal. The following examples help to show this concept.

Example 1: The fraction ½ is equal to 1 divided by 2, which is .5.

Example 2: The fraction ⁵⁄₁₆ is equal to 5 divided by 16, which is .3125.

Another way to think of a common fraction is that the denominator (bottom number) is the denomination or type of unit, while the numerator (top number) is the amount of that particular type of unit. For example, in the common fraction ¾, the denomination is quarters, and 3 represents how many quarters you have.

Reducing Fractions

When working with fractions, it is customary to reduce fractional computations to their lowest form. A fraction such as ⅛ is considered to be in lowest form, whereas ⅜ is not. This is because ⅜ can be reduced to ¼. The method used to reduce a fraction to its lowest form is to find the largest number that both the numerator and denominator are divisible by and then divide them both, as shown in the examples below.

Example 1: Reduce the fraction ⁸⁄₃₂ to its lowest form.

Because 8 is the largest number that 8 and 32 are both divisible by, we divide them both by 8 to get our answer of ¼.

$$\frac{8 \div 8}{32 \div 8} = \frac{1}{4}$$

Example 2: Reduce the fraction ¹⁴⁄₁₆ to its lowest form.

Because 2 is the largest number that 14 and 16 are both divisible by, we divide them both by 2 to get our answer of ⅞

$$\frac{14 \div 2}{16 \div 2} = \frac{7}{8}$$

Math Supplement Practice Exercise 1

Determine which number is the numerator and which number is the denominator.

	Fraction	Numerator	Denominator
1.	$\dfrac{3}{8}$		
2.	$\dfrac{7}{16}$		
3.	$\dfrac{23}{32}$		
4.	$\dfrac{12}{8}$		

Change the following common fractions into their decimal equivalents.

	Fraction	Decimal Equivalent
5.	$\dfrac{3}{8}$	
6.	$\dfrac{7}{16}$	
7.	$\dfrac{23}{32}$	
8.	$\dfrac{12}{8}$	

Write the following fractions in their lowest form.

	Fraction	Lowest Form
9.	$\dfrac{4}{16}$	
10.	$\dfrac{16}{64}$	
11.	$\dfrac{24}{32}$	
12.	$\dfrac{2}{8}$	
13.	$\dfrac{6}{32}$	
14.	$\dfrac{13}{32}$	

Notes and Specifications

OBJECTIVES

- Be able to identify notes and specifications
- Be able to identify different hole types
- Be able to interpret thread specifications for holes

KEY TERMS

General note	Through hole	Countersink
Local note	Blind hole	Counterdrill
Flag note	Spotface	Tap
Specifications	Counterbore	

OVERVIEW

Notes and specifications are included on drawings to provide additional information needed to understand and properly manufacture the item(s) shown on the print.

NOTES

Notes can be divided into three categories, general notes, local notes, and flag notes (see Figure 5-1). A **general note** applies to the entire drawing or set of drawings, a **local note** applies only to a localized part of the drawing, and a **flag note** applies anywhere the specific flag is shown. Flag notes are basically local notes that have a flag placed at the applicable area and the note placed elsewhere on the drawing.

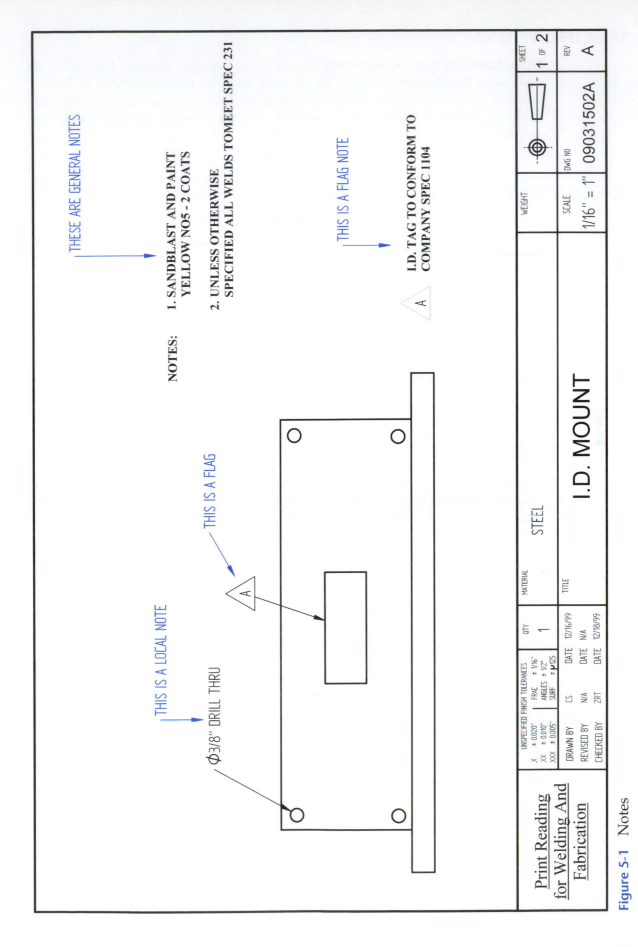

Figure 5-1 Notes

SPECIFICATIONS

Specifications may be shown on drawings as general, local, or flag notes, or they may be shown within a block area such as the title block or bill of materials. Specifications are given for materials, processing requirements, welding and heat-treating requirements, threaded part requirements, and so on. Several examples are shown in Figure 5-2.

SPECIFICATIONS AND DIMENSIONS FOR HOLES

A hole can go all the way through the part or partially through the part. When a hole goes all the way through, it is called a **through hole**; when it goes partially through, it is called a **blind hole**. Through holes may be made by many different processes (drilling, oxy-fuel, plasma, laser, and so on). They are dimensioned by the diameter, usually by a leader, as shown in Figure 5-3 or with extension and dimension lines, as described in Chapter 4. The diameter symbol or the abbreviation DIA is used to denote diameter. THRU is used to note that the hole is to go all the way through the part. Through holes that are made by drilling or machining are typically dimensioned using a precision decimal dimension, whereas holes made by oxy-fuel, arc cutting, or other nonprecision processes are typically dimensioned using common fractions. Blind holes are dimensioned as shown in Figure 5-3, with the diameter and the depth of the hole not including the drill point. The depth is indicated by the abbreviation DP or by the symbol with the line and downward arrow, as shown.

Both through holes and blind holes may have added features such as: **spotface**, **counterbore**, **countersink**, **counterdrill**, and **tap**. Each of these will be discussed in the following chapter sections.

SPOTFACE

Spotfacing is a machining process that removes a small amount of material from an area around the surface of a hole to a diameter and depth that will allow a bolt head, nut, or washer to seat properly against the base metal when it is tightened. See Figure 5-4.

COUNTERBORE

The counterbore feature removes material to a specified depth and diameter around the surface of a hole to allow the head of a bolt to be recessed to the specified depth. See Figure 5-5.

COUNTERSINK

The countersink feature removes material from the surface of a hole at a specified angle so that flathead-type screws will set flush with the surface of the part. See Figure 5-6.

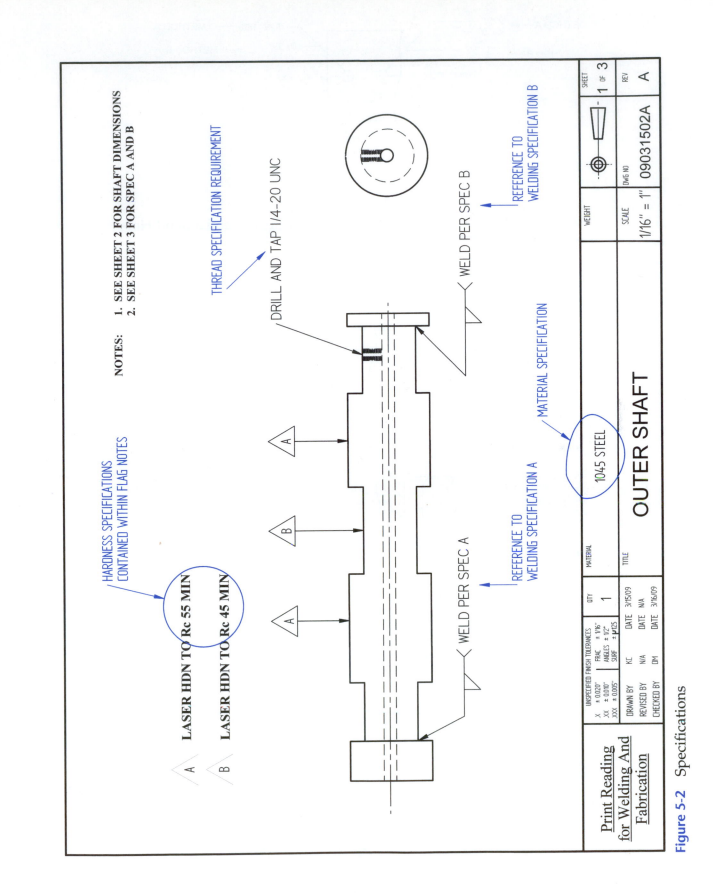

Figure 5-2 Specifications

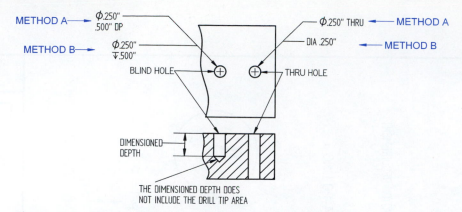

Figure 5-3   Methods of Dimensioning a Through Hole and a Blind Hole

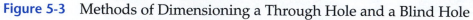

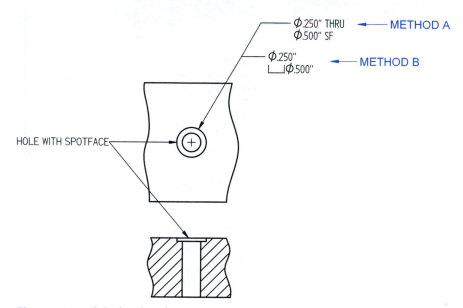

Figure 5-4   Methods of Dimensioning a Spotface Hole

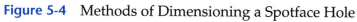

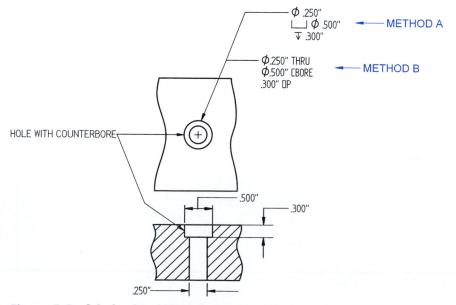

Figure 5-5   Methods of Dimensioning a Counterbore Hole

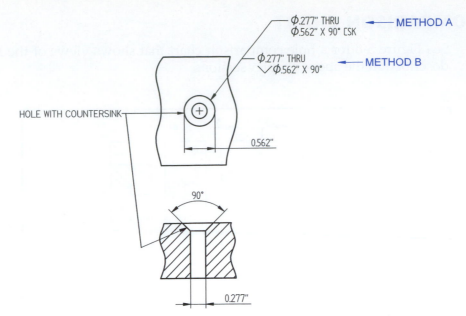

Figure 5-6   Methods of Dimensioning a Countersink Hole

COUNTERDRILL

A counterdrilled hole allows a flathead-type screw to be recessed to a specified depth below the surface of the part. See Figure 5-7.

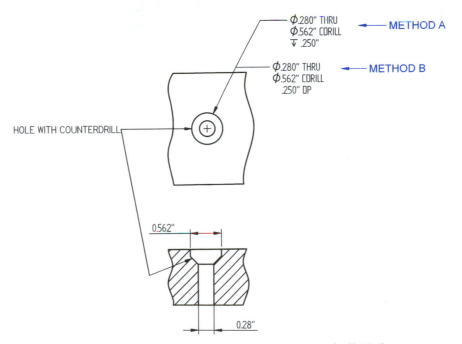

Figure 5-7   Methods of Dimensioning a Counterdrill Hole

HOLE COMPARISON CHART

See Figure 5-8 for a hole comparison chart that shows views of the hole types described in the above chapter sections.

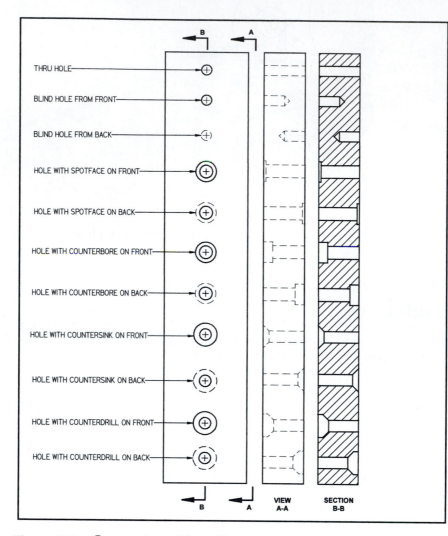

Figure 5-8 Comparison Chart Showing Different Hole Preparation Types

THREAD SPECIFICATIONS

Assembly drawings that welders work with may include bolted connections and tapped holes. The basic specifications used for threaded bolts and tapped holes that are most useful to the welder or fabricator designate the nominal size, pitch, and form of thread. Additional information such as class of fit, right-hand or left-hand direction of the threads, internal or external threads, and thread depth (DP) may be included as required. Figure 5-9 and Figure 5-10 show how the typical basic thread designations for both U.S. and metric threads are shown on drawings.

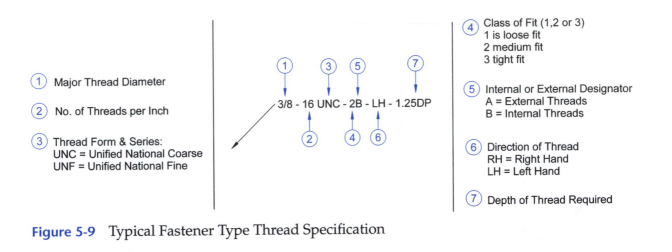

Figure 5-9   Typical Fastener Type Thread Specification

Figure 5-10   Typical Basic Metric Thread Specification

Practice Exercise 1

Define the following key terms.

1. General note:

2. Local note:

3. Flag note:

4. Specification:

5. Through hole:

6. Blind hole:

7. Tap:

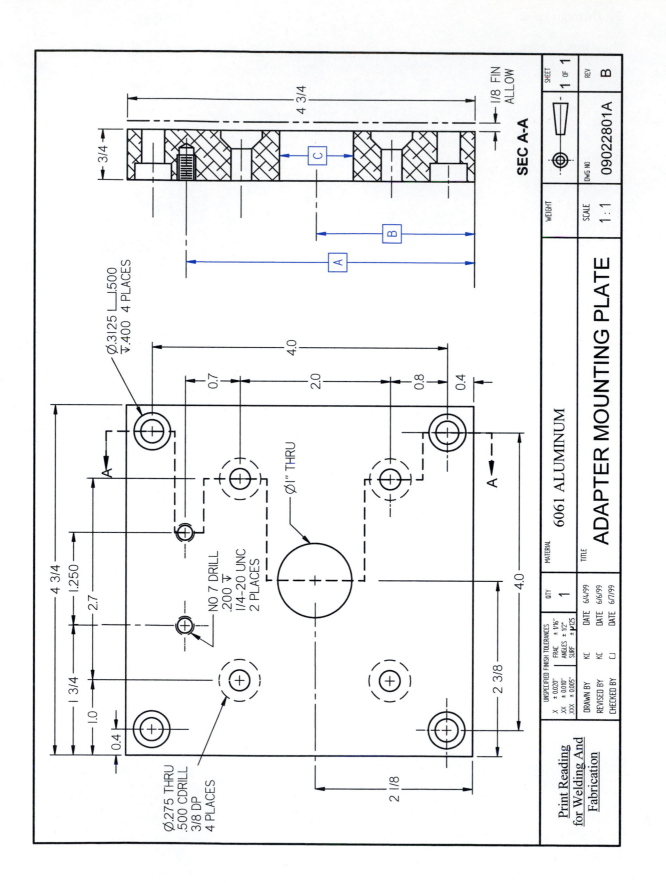

ADAPTER MOUNTING PLATE

UNSPECIFIED FINISH TOLERANCES			QTY	MATERIAL	
X ± 0.020"	FRAC ± 1/16"		1	6061 ALUMINUM	
XX ± 0.010"	ANGLES ± 1/2"				
XXX ± 0.005"	SURF ± μ125				
DRAWN BY	KC	DATE 6/4/99			TITLE
REVISED BY	KC	DATE 6/6/99			
CHECKED BY	CJ	DATE 6/7/99			

WEIGHT		SHEET 1 OF 1
	SCALE 1:1	REV B
DWG NO 09022801A		

Print Reading for Welding And Fabrication

SEC A-A

4 3/4

3/4

1/8 FIN ALLOW

Ø.3125 ⌴.500 ⌵.400 4 PLACES

4.0

0.7

2.0

0.8

0.4

Ø1" THRU

NO 7 DRILL .200 ⌵ 1/4-20 UNC 2 PLACES

4 3/4

1.250

2.7

1 3/4

1.0

0.4

4.0

2 3/8

2 1/8

Ø.275 THRU .500 CDRILL 3/8 DP 4 PLACES

Practice Exercise 2

Answer the following questions. Refer to drawing number 09022801A REV B and the drawing below.

1. What is distance A? _____

2. What is distance B? _____

3. What is distance C? _____

4. What is distance D? _____

5. What is distance E? _____

6. What is distance F? _____

7. What is distance G? _____

8. What is distance H? _____

9. What is distance I? _____

10. What does CDRILL stand for? _____

11. Is this drawing first-angle or third-angle projection? _____

12. What is the specification for the tap that is required to complete this part? _____

13. How deep is the counterbore? _____

Math Supplement

Fractions with Common Denominators

Addition of Fractions with Common Denominators

When adding fractions that have like denominators, use the following property:

$$\frac{a}{c} + \frac{b}{c} = \frac{a+b}{c}$$

Example 1:

$$\frac{3}{8} + \frac{5}{8} = \frac{3+5}{8} = \frac{8}{8}$$

Example 2:

$$\frac{7}{16} + \frac{5}{16} = \frac{7+5}{16} = \frac{12}{16}$$

Subtraction of Fractions with Common Denominators

When subtracting fractions that have like denominators, use the following property:

$$\frac{a}{c} - \frac{b}{c} = \frac{a-b}{c}$$

Example 3:

$$\frac{15}{16} - \frac{5}{16} = \frac{15-5}{16} = \frac{10}{16} = \frac{5}{8}$$

CHAPTER 5

Math Supplement Practice Exercise 1

Use addition or subtraction to compute the following. Write your answers in lowest form.

1. $\dfrac{5}{8} + \dfrac{1}{8} =$

2. $\dfrac{7}{32} - \dfrac{4}{32} =$

3. $\dfrac{23}{64} - \dfrac{17}{64} =$

4. $\dfrac{3}{4} - \dfrac{1}{4} =$

5. $\dfrac{17}{32} + \dfrac{8}{32} =$

6. $\dfrac{13}{32} - \dfrac{7}{32} =$

7. $\dfrac{3}{8} + \dfrac{3}{8} =$

8. $\dfrac{1}{16} + \dfrac{3}{16} =$

9. $\dfrac{19}{32} - \dfrac{3}{32} =$

10. $\dfrac{5}{32} + \dfrac{9}{32} =$

11. $\dfrac{29}{64} - \dfrac{7}{64} =$

12. $\dfrac{7}{8} + \dfrac{1}{8} =$

13. $\dfrac{5}{16} - \dfrac{1}{16} =$

14. $\dfrac{3}{8} + \dfrac{1}{8} =$

15. $\dfrac{3}{16} + \dfrac{5}{16} =$

16. $\dfrac{13}{32} - \dfrac{5}{32} =$

17. $\dfrac{3}{16} - \dfrac{1}{16} =$

18. $\dfrac{3}{8} - \dfrac{1}{8} =$

Materials

OBJECTIVES

- Understand the importance of identifying and using the required metal for the job
- Have a basic understanding of how industry classifies metals
- Be able to identify and name the typical standard shapes used in industry
- Understand how the different materials are described in a bill of materials list

KEY TERMS

AISI-SAE Steel Classification System

Unified Numbering System

Sheet-Metal

Strip or Band

Plate

Bar—Flat, Square, Hexagon, Octagon, Round, Half Oval

Angle

Channel

Zee

HP—Bearing Pile Beam

W—Wide-flange Beam

S—Standard Beam

M—Miscellaneous Beam

Tee

Tube—Square and Rectangular

Round Tubing

Pipe

OVERVIEW

Welded components used in fabrication, structures, piping systems, and so on, use many types of stock and custom materials that can be purchased in different shapes and sizes. Each of the materials required for the design shown on a drawing can be found in the different views, and each one is specified by type and size in an area of the title block, bill of materials, or notes. The decision about the placement of material specifications on the drawing depends mostly on the quantity of materials required. For example, a drawing of a simple, one-piece bracket, where only one piece of material is required, might have the material type and size specified within a space in the title block or in a note on the drawing. A multipiece assembly would normally use a bill of material to list all of the materials, with their sizes and specifications. As discussed in Chapter 1 of this text, the item number, quantity, size, part number, and material type may all be listed to give the description of the material.

METALS

Many types of metals such as aluminum, steel, magnesium, nickel, and so on, and alloys such as aluminum-copper alloys, nickel-chrome alloys, nickel-copper, and others can be used in welding and fabrication. When working from a drawing, it is extremely important for the welder or fabricator to use the specific metal required, as stated on the drawing and/or job requirements. The engineer determines the type of metal to use based on matching properties of the metal with the properties required for proper design. Some of the properties of metals that engineers are concerned with include strength, ductility, hardness, toughness, fatigue strength, corrosion resistance, abrasion resistance, and impact resistance. Many metals look the same and cannot be identified without some type of identification system. For example, two materials that look similar are AISI 1018 and AISI 1045. Both are carbon steels; however, 1018 has approximately .18 percent carbon, while 1045 has approximately .45 percent carbon. The difference in carbon between these two steels has an effect on the strength, hardenability, and crack sensitivity of the metal. Identification systems are used to ensure that the correct metal is used for the job. Each company has its own method for marking and identifying metal. Some use a color-code system; others use numbering systems of differing types. Whatever type is used, the welder or fabricator needs to know the system, be able to identify the type of metal as listed on the drawing and/or bill of material, and ensure that the correct metal is used.

INDUSTRY IDENTIFICATION SYSTEMS FOR METALS

As discussed above, each individual company should have a method for ensuring that metals are properly identified. Ultimately, the particular company identification method correlates to one of a number of industry classification numbering systems or at times particular producers' product or trade names. Some of the numbering systems include those developed by the U.S. military, the American Iron and Steel Institute (AISI), the Society of Automotive Engineers (SAE), the American Society of Mechanical Engineers (ASME), the American National Standards Institute (ANSI), and others. Many charts have been developed that can be used to reference and cross-reference metals to their numbering systems. A common system used for steel is the **AISI-SAE Steel Classification System**. This system uses four or five numbers to classify specific types of steel. Table 6-1 shows the basics of the AISI-SAE system, the different groups of metals it covers, and how it works.

Table 6-2 shows the basics of a system developed by the American Society of Testing Materials (ASTM) and the Society of Automotive Engineers (SAE). This system, called the **Unified Numbering System (UNS)** uses a letter to designate the general type of metal, and five numbers to identify the specific metal within the general metal type.

Table 6-1   AISI-SAE Steel Classifications

X X X X X

The last two or three numbers give the nominal carbon content in the steel, in hundreths of a percent

The first 2 numbers designate the type of steel or steel alloy

AISI – SAE Number	Steel Group
10XX	Carbon Steels
11XX	
12XX	
15XX	
13XX	Manganese Steels
23XX	Nickel Steels
25XX	
31XX	Nickel – Chromium Steels
32XX	
33XX	
34XX	
40XX	Molybdenum Steels
44XX	
41XX	Chromium - Molybdenum Steels
43XX	Nickel – Chromium – Molybdenum Steels
47XX	
81XX	
86XX	
87XX	
88XX	
93XX	
94XX	
97XX	
98XX	
46XX	Nickel – Molybdenum Steels
48XX	
50XXX	Chromium Steels
51XXX	
52XXX	
61XX	Chromium – Vanadium Steels
72XX	Tungsten - Chromium Steels
92XX	Silicon Manganese Steels

Note: 1. XX represents the approximate carbon content of the steel in hundredths of a percent.
2. This chart is for educational purposes only and should not be used for design, fabrication, manufacturing, etc.
3. For the most recent & complete AISI-SAE Steel Classifications please contact either AISI or SAE.

Table 6-2 UNS Numbering

X00001 to X99999

The Last 5 Numbers Designate the Specific Metal Within the Metal Type

Letter Designator Used to Specify Metal Type

UNS Letter Designator	Metal
A	Aluminum and Aluminum Alloys
C	Copper and Copper Alloys
D	Specified Mechanical Property Steels
E	Rare Earth and Rare Earthlike Metals & Alloys
F	Cast Iron
G	AISI-SAE Carbon Steels and Carbon Alloy Steels (Not Tool Steels)
H	AISI and SAE H-Steels
J	Cast Steels with the exception of Tool Steels
K	Miscellaneous Steels & Ferrous Alloys
L	Low Melting Point Metals and Alloys
M	Miscellaneous Metals and Alloys
N	Nickel and Nickel Alloys
P	Precious Metals and Alloys
R	Reactive and Refractory Metals and Alloys
S	Heat and Corrosion Resistant Stainless Steels
T	Tool Steels (Wrought and Cast)
W	Welding Filler Metals
Z	Zinc and Zinc Alloys

Note: 1. This chart is for educational purposes only and should not be used for design, fabrication, manufacturing, etc.
2. For the most recent & complete UNS Metal Classifications please contact either ASTM or SAE.

Each of the different types of metals may come in different shapes and sizes. Figures 6-1 through 6-21 on the following pages show the typical standard material shapes sold by suppliers and used by manufacturers and constructors in industry, along with the typical methods of size description for each.

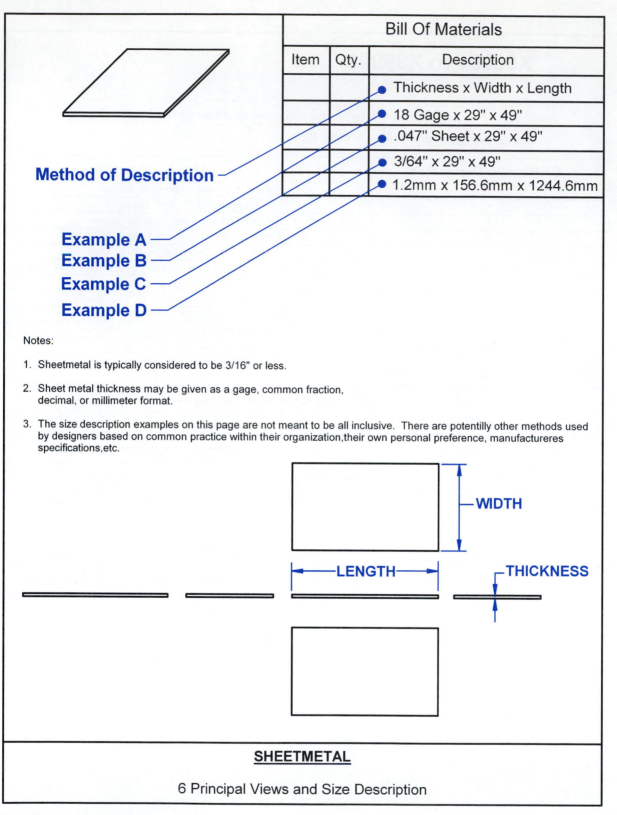

Bill Of Materials		
Item	Qty.	Description
		Thickness x Width x Length
		18 Gage x 29" x 49"
		.047" Sheet x 29" x 49"
		3/64" x 29" x 49"
		1.2mm x 156.6mm x 1244.6mm

Method of Description

Example A
Example B
Example C
Example D

Notes:

1. Sheetmetal is typically considered to be 3/16" or less.

2. Sheet metal thickness may be given as a gage, common fraction, decimal, or millimeter format.

3. The size description examples on this page are not meant to be all inclusive. There are potentilly other methods used by designers based on common practice within their organization,their own personal preference, manufactureres specifications,etc.

WIDTH

LENGTH

THICKNESS

SHEETMETAL

6 Principal Views and Size Description

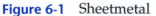
Figure 6-1 Sheetmetal

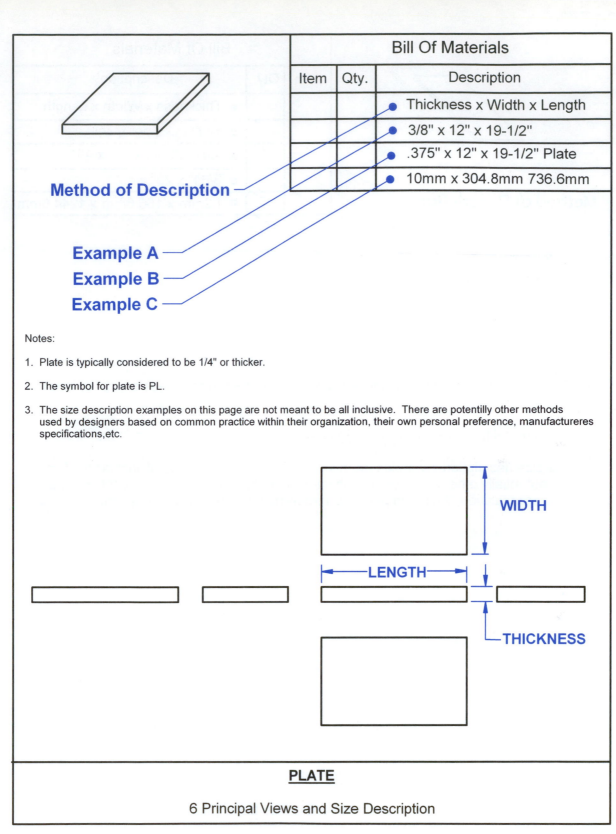

Bill Of Materials		
Item	Qty.	Description
		Thickness x Width x Length
		3/8" x 12" x 19-1/2"
		.375" x 12" x 19-1/2" Plate
		10mm x 304.8mm 736.6mm

Method of Description

Example A

Example B

Example C

Notes:

1. Plate is typically considered to be 1/4" or thicker.

2. The symbol for plate is PL.

3. The size description examples on this page are not meant to be all inclusive. There are potentilly other methods used by designers based on common practice within their organization, their own personal preference, manufactureres specifications,etc.

WIDTH

LENGTH

THICKNESS

PLATE

6 Principal Views and Size Description

Figure 6-2   Plate

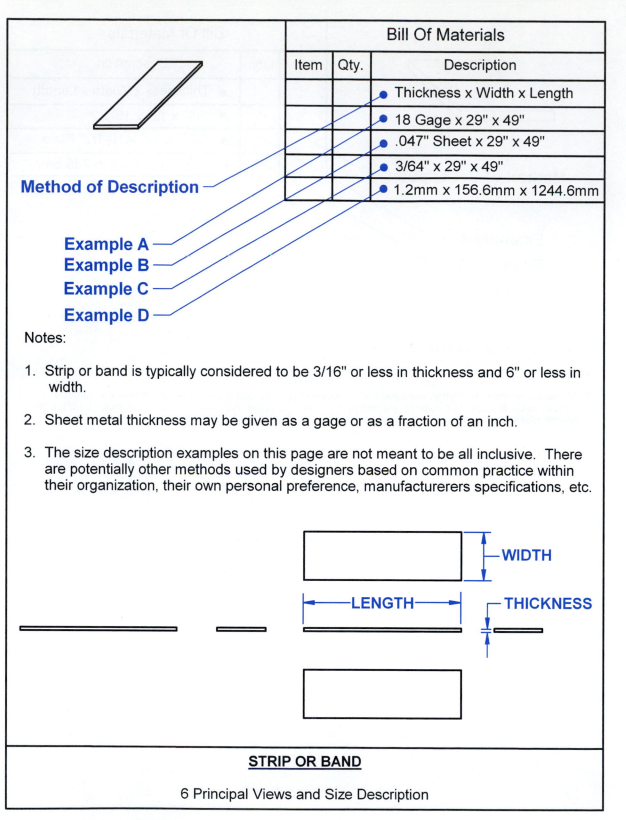

Bill Of Materials		
Item	Qty.	Description
		Thickness x Width x Length
		18 Gage x 29" x 49"
		.047" Sheet x 29" x 49"
		3/64" x 29" x 49"
		1.2mm x 156.6mm x 1244.6mm

Method of Description

Example A
Example B
Example C
Example D

Notes:

1. Strip or band is typically considered to be 3/16" or less in thickness and 6" or less in width.

2. Sheet metal thickness may be given as a gage or as a fraction of an inch.

3. The size description examples on this page are not meant to be all inclusive. There are potentially other methods used by designers based on common practice within their organization, their own personal preference, manufacturerers specifications, etc.

WIDTH

LENGTH   THICKNESS

STRIP OR BAND

6 Principal Views and Size Description

Figure 6-3   Strip or Band

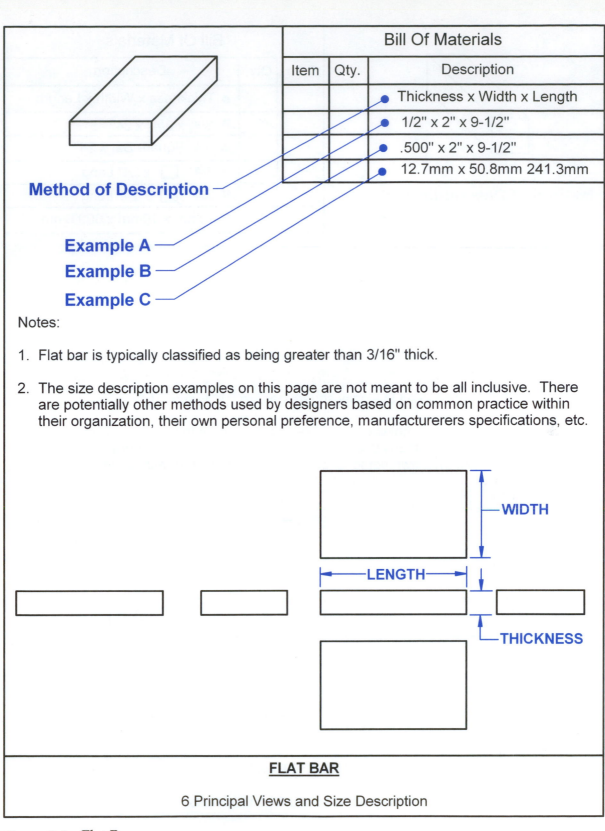

Bill Of Materials		
Item	Qty.	Description
		Thickness x Width x Length
		1/2" x 2" x 9-1/2"
		.500" x 2" x 9-1/2"
		12.7mm x 50.8mm 241.3mm

Method of Description

Example A

Example B

Example C

Notes:

1. Flat bar is typically classified as being greater than 3/16" thick.

2. The size description examples on this page are not meant to be all inclusive. There are potentially other methods used by designers based on common practice within their organization, their own personal preference, manufacturerers specifications, etc.

WIDTH

LENGTH

THICKNESS

FLAT BAR

6 Principal Views and Size Description

Figure 6-4   Flat Bar

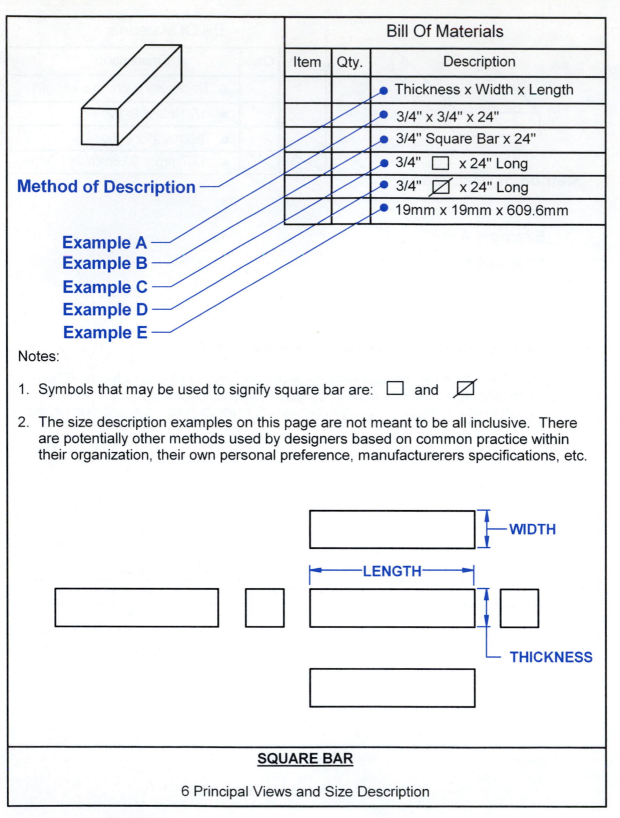

Bill Of Materials		
Item	Qty.	Description
		Thickness x Width x Length
		3/4" x 3/4" x 24"
		3/4" Square Bar x 24"
		3/4" □ x 24" Long
		3/4" ⊠ x 24" Long
		19mm x 19mm x 609.6mm

Method of Description

Example A
Example B
Example C
Example D
Example E

Notes:

1. Symbols that may be used to signify square bar are: □ and ⊠

2. The size description examples on this page are not meant to be all inclusive. There are potentially other methods used by designers based on common practice within their organization, their own personal preference, manufacturerers specifications, etc.

WIDTH

LENGTH

THICKNESS

SQUARE BAR

6 Principal Views and Size Description

Figure 6-5 Square Bar

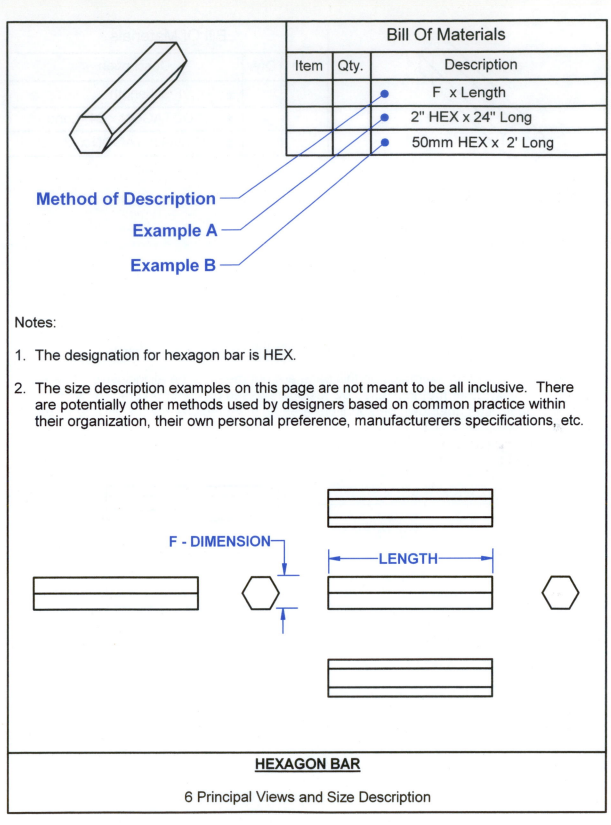

Bill Of Materials		
Item	Qty.	Description
		F x Length
		2" HEX x 24" Long
		50mm HEX x 2' Long

Method of Description

Example A

Example B

Notes:

1. The designation for hexagon bar is HEX.

2. The size description examples on this page are not meant to be all inclusive. There are potentially other methods used by designers based on common practice within their organization, their own personal preference, manufacturerers specifications, etc.

F - DIMENSION

LENGTH

HEXAGON BAR

6 Principal Views and Size Description

Figure 6-6   Hexagon Bar

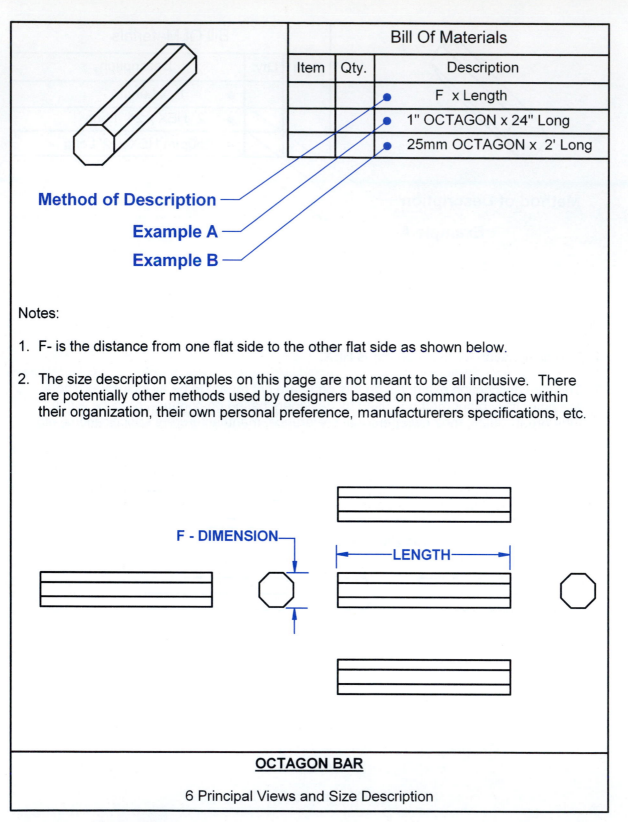

Bill Of Materials

Item	Qty.	Description
		F x Length
		1" OCTAGON x 24" Long
		25mm OCTAGON x 2' Long

Method of Description

Example A

Example B

Notes:

1. F- is the distance from one flat side to the other flat side as shown below.

2. The size description examples on this page are not meant to be all inclusive. There are potentially other methods used by designers based on common practice within their organization, their own personal preference, manufacturerers specifications, etc.

F - DIMENSION

LENGTH

OCTAGON BAR

6 Principal Views and Size Description

Figure 6-7 Octagon Bar

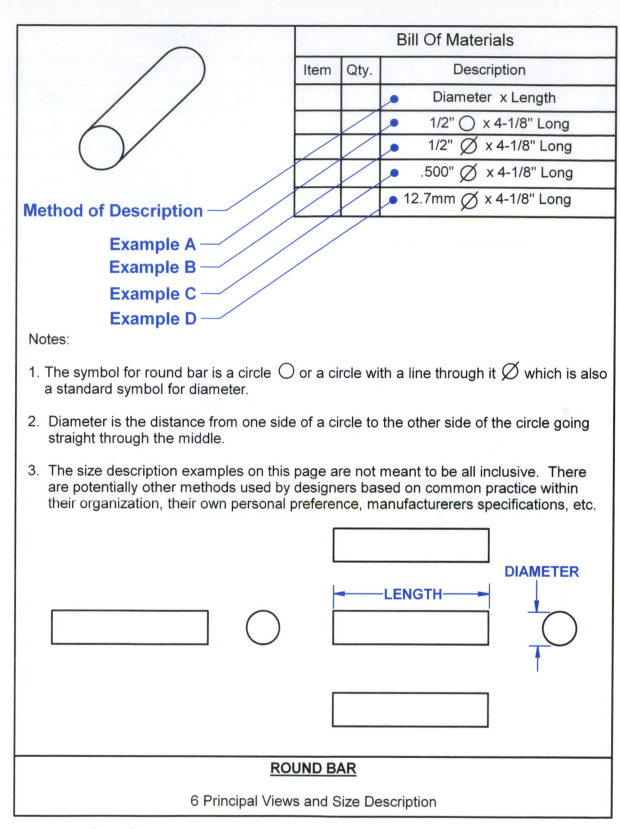

Bill Of Materials		
Item	Qty.	Description
		Diameter x Length
		1/2" ◯ x 4-1/8" Long
		1/2" ⌀ x 4-1/8" Long
		.500" ⌀ x 4-1/8" Long
		12.7mm ⌀ x 4-1/8" Long

Method of Description

Example A

Example B

Example C

Example D

Notes:

1. The symbol for round bar is a circle ◯ or a circle with a line through it ⌀ which is also a standard symbol for diameter.

2. Diameter is the distance from one side of a circle to the other side of the circle going straight through the middle.

3. The size description examples on this page are not meant to be all inclusive. There are potentially other methods used by designers based on common practice within their organization, their own personal preference, manufacturerers specifications, etc.

DIAMETER

←—LENGTH—→

ROUND BAR

6 Principal Views and Size Description

Figure 6-8   Round Bar

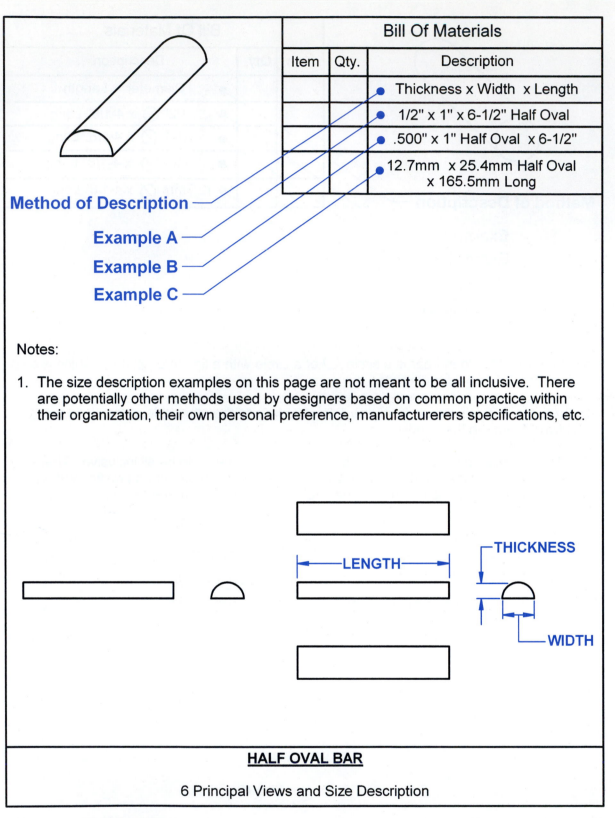

Bill Of Materials		
Item	Qty.	Description
		Thickness x Width x Length
		1/2" x 1" x 6-1/2" Half Oval
		.500" x 1" Half Oval x 6-1/2"
		12.7mm x 25.4mm Half Oval x 165.5mm Long

Method of Description

Example A

Example B

Example C

Notes:

1. The size description examples on this page are not meant to be all inclusive. There are potentially other methods used by designers based on common practice within their organization, their own personal preference, manufacturerers specifications, etc.

LENGTH

THICKNESS

WIDTH

HALF OVAL BAR

6 Principal Views and Size Description

Figure 6-9   Half Oval Bar

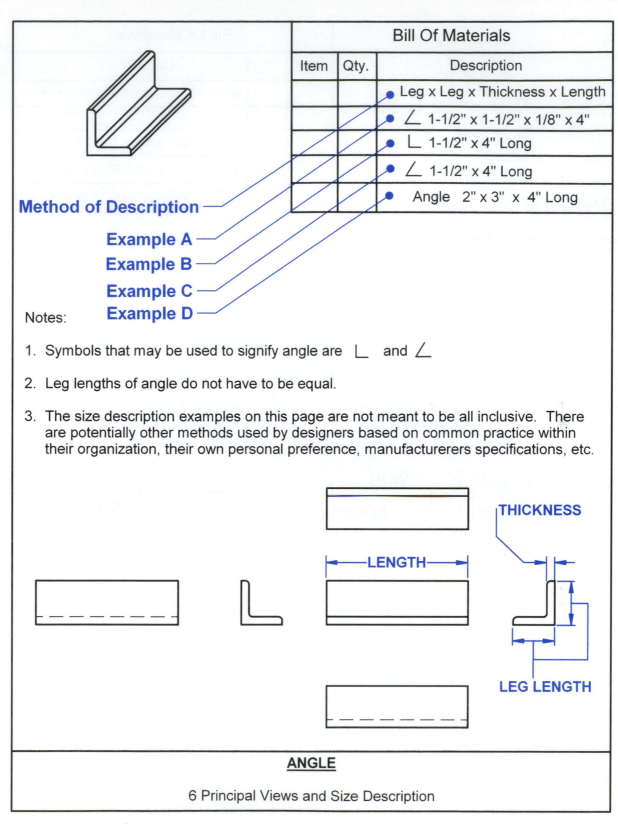

Bill Of Materials		
Item	Qty.	Description
		Leg x Leg x Thickness x Length
		∠ 1-1/2" x 1-1/2" x 1/8" x 4"
		∟ 1-1/2" x 4" Long
		∠ 1-1/2" x 4" Long
		Angle 2" x 3" x 4" Long

Method of Description

Example A

Example B

Example C

Example D

Notes:

1. Symbols that may be used to signify angle are ∟ and ∠

2. Leg lengths of angle do not have to be equal.

3. The size description examples on this page are not meant to be all inclusive. There are potentially other methods used by designers based on common practice within their organization, their own personal preference, manufacturerers specifications, etc.

THICKNESS

LENGTH

LEG LENGTH

ANGLE

6 Principal Views and Size Description

Figure 6-10 Angle

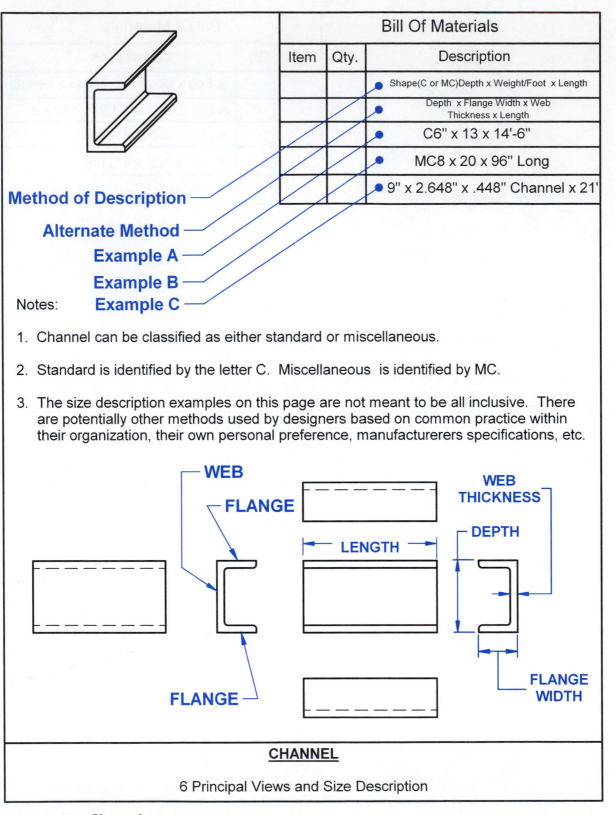

Bill Of Materials		
Item	Qty.	Description
		Shape(C or MC)Depth x Weight/Foot x Length
		Depth x Flange Width x Web Thickness x Length
		C6" x 13 x 14'-6"
		MC8 x 20 x 96" Long
		9" x 2.648" x .448" Channel x 21'

Method of Description

Alternate Method

Example A

Example B

Notes: **Example C**

1. Channel can be classified as either standard or miscellaneous.

2. Standard is identified by the letter C. Miscellaneous is identified by MC.

3. The size description examples on this page are not meant to be all inclusive. There are potentially other methods used by designers based on common practice within their organization, their own personal preference, manufacturerers specifications, etc.

WEB

FLANGE

WEB THICKNESS

LENGTH

DEPTH

FLANGE

FLANGE WIDTH

CHANNEL

6 Principal Views and Size Description

Figure 6-11   Channel

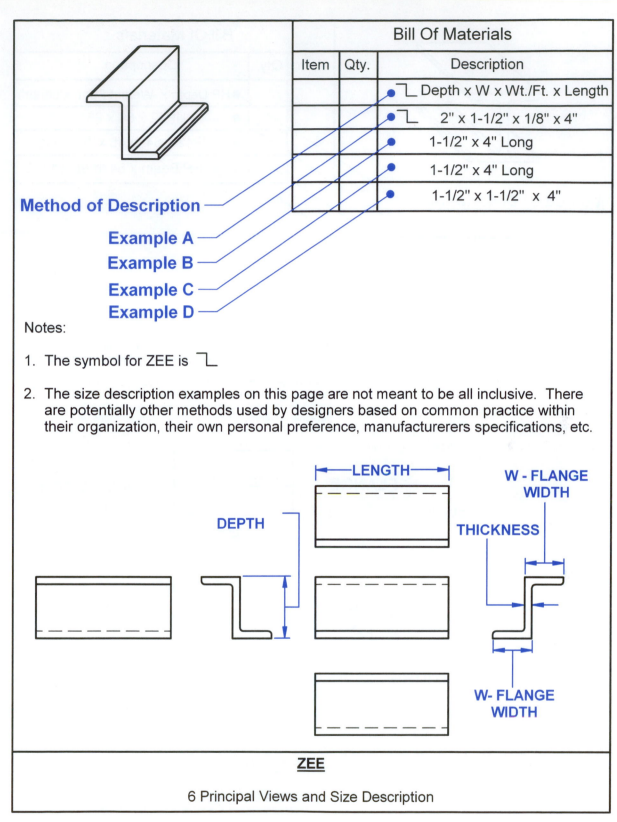

Bill Of Materials

Item	Qty.	Description
		⌐L Depth x W x Wt./Ft. x Length
		⌐L 2" x 1-1/2" x 1/8" x 4"
		1-1/2" x 4" Long
		1-1/2" x 4" Long
		1-1/2" x 1-1/2" x 4"

Method of Description

Example A

Example B

Example C

Example D

Notes:

1. The symbol for ZEE is ⌐L

2. The size description examples on this page are not meant to be all inclusive. There are potentially other methods used by designers based on common practice within their organization, their own personal preference, manufacturerers specifications, etc.

LENGTH

DEPTH

W - FLANGE WIDTH

THICKNESS

W- FLANGE WIDTH

ZEE

6 Principal Views and Size Description

Figure 6-12 ZEE

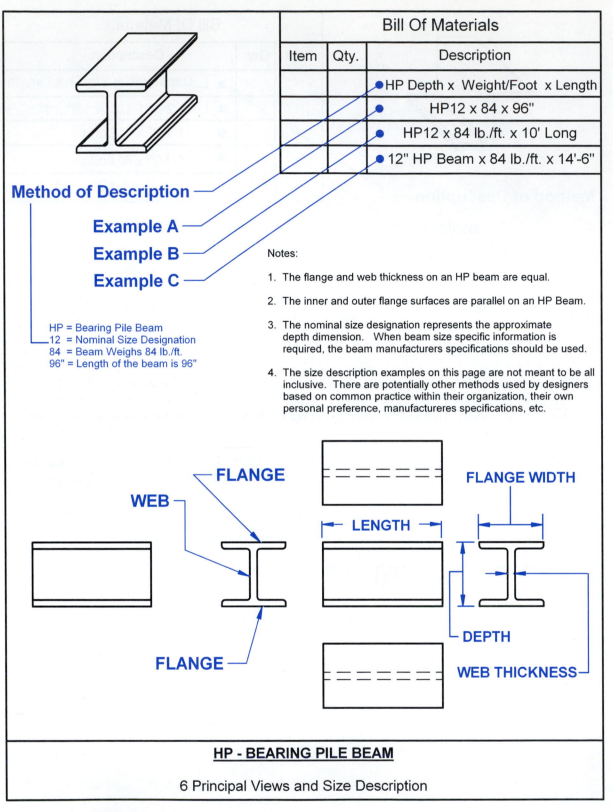

Bill Of Materials

Item	Qty.	Description
		HP Depth x Weight/Foot x Length
		HP12 x 84 x 96"
		HP12 x 84 lb./ft. x 10' Long
		12" HP Beam x 84 lb./ft. x 14'-6"

Method of Description

Example A

Example B

Example C

HP = Bearing Pile Beam
12 = Nominal Size Designation
84 = Beam Weighs 84 lb./ft.
96" = Length of the beam is 96"

Notes:

1. The flange and web thickness on an HP beam are equal.

2. The inner and outer flange surfaces are parallel on an HP Beam.

3. The nominal size designation represents the approximate depth dimension. When beam size specific information is required, the beam manufacturers specifications should be used.

4. The size description examples on this page are not meant to be all inclusive. There are potentially other methods used by designers based on common practice within their organization, their own personal preference, manufactureres specifications, etc.

WEB

FLANGE

FLANGE WIDTH

LENGTH

FLANGE

DEPTH

WEB THICKNESS

HP - BEARING PILE BEAM

6 Principal Views and Size Description

Figure 6-13 HP Beam

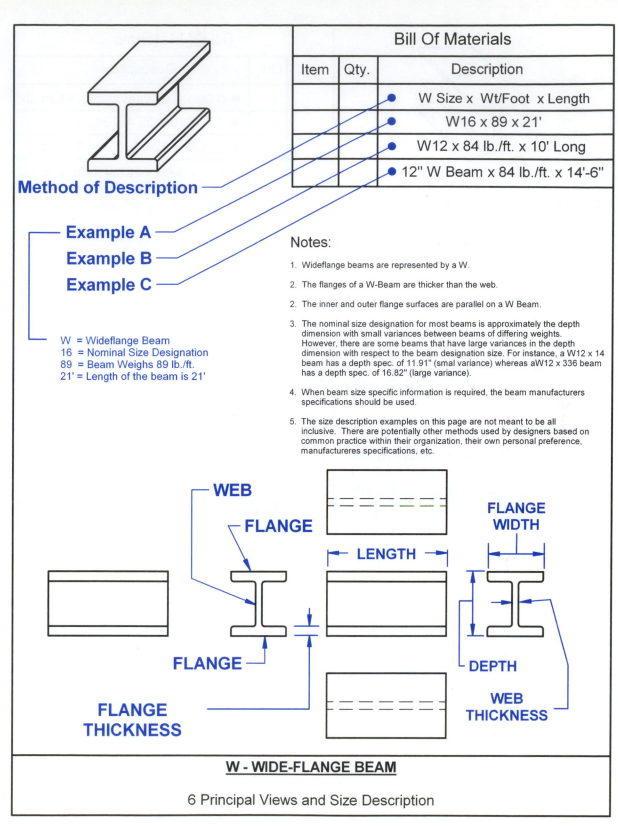

Method of Description

Example A
Example B
Example C

W = Wideflange Beam
16 = Nominal Size Designation
89 = Beam Weighs 89 lb./ft.
21' = Length of the beam is 21'

Bill Of Materials		
Item	Qty.	Description
		W Size x Wt/Foot x Length
		W16 x 89 x 21'
		W12 x 84 lb./ft. x 10' Long
		12" W Beam x 84 lb./ft. x 14'-6"

Notes:

1. Wideflange beams are represented by a W.

2. The flanges of a W-Beam are thicker than the web.

2. The inner and outer flange surfaces are parallel on a W Beam.

3. The nominal size designation for most beams is approximately the depth dimension with small variances between beams of differing weights. However, there are some beams that have large variances in the depth dimension with respect to the beam designation size. For instance, a W12 x 14 beam has a depth spec. of 11.91" (smal variance) whereas aW12 x 336 beam has a depth spec. of 16.82" (large variance).

4. When beam size specific information is required, the beam manufacturers specifications should be used.

5. The size description examples on this page are not meant to be all inclusive. There are potentially other methods used by designers based on common practice within their organization, their own personal preference, manufactureres specifications, etc.

WEB

FLANGE

FLANGE

FLANGE THICKNESS

LENGTH

FLANGE WIDTH

DEPTH

WEB THICKNESS

W - WIDE-FLANGE BEAM

6 Principal Views and Size Description

Figure 6-14 Wide-Flange Beam

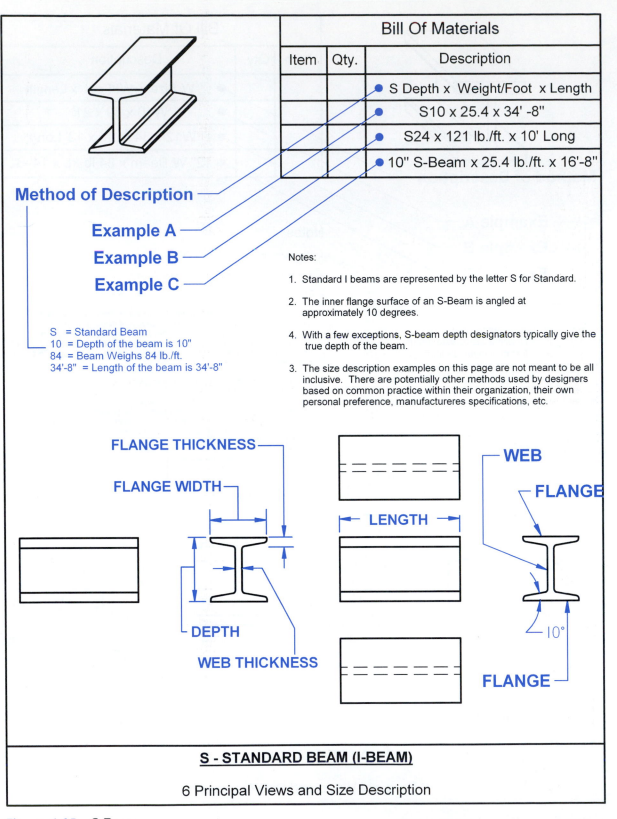

Bill Of Materials

Item	Qty.	Description
		S Depth x Weight/Foot x Length
		S10 x 25.4 x 34' -8"
		S24 x 121 lb./ft. x 10' Long
		10" S-Beam x 25.4 lb./ft. x 16'-8"

Method of Description

Example A

Example B

Example C

S = Standard Beam
10 = Depth of the beam is 10"
84 = Beam Weighs 84 lb./ft.
34'-8" = Length of the beam is 34'-8"

Notes:

1. Standard I beams are represented by the letter S for Standard.

2. The inner flange surface of an S-Beam is angled at approximately 10 degrees.

4. With a few exceptions, S-beam depth designators typically give the true depth of the beam.

3. The size description examples on this page are not meant to be all inclusive. There are potentially other methods used by designers based on common practice within their organization, their own personal preference, manufactureres specifications, etc.

FLANGE THICKNESS

FLANGE WIDTH

WEB

FLANGE

LENGTH

DEPTH

WEB THICKNESS

10°

FLANGE

S - STANDARD BEAM (I-BEAM)

6 Principal Views and Size Description

Figure 6-15   S-Beam

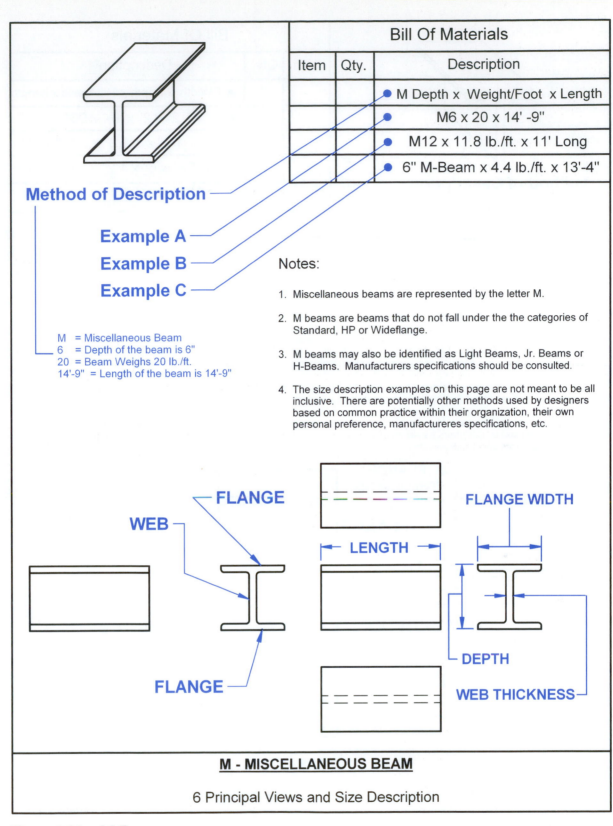

Bill Of Materials

Item	Qty.	Description
		M Depth x Weight/Foot x Length
		M6 x 20 x 14' -9"
		M12 x 11.8 lb./ft. x 11' Long
		6" M-Beam x 4.4 lb./ft. x 13'-4"

Method of Description

Example A

Example B

Example C

M = Miscellaneous Beam
6 = Depth of the beam is 6"
20 = Beam Weighs 20 lb./ft.
14'-9" = Length of the beam is 14'-9"

Notes:

1. Miscellaneous beams are represented by the letter M.

2. M beams are beams that do not fall under the the categories of Standard, HP or Wideflange.

3. M beams may also be identified as Light Beams, Jr. Beams or H-Beams. Manufacturers specifications should be consulted.

4. The size description examples on this page are not meant to be all inclusive. There are potentially other methods used by designers based on common practice within their organization, their own personal preference, manufactureres specifications, etc.

FLANGE

WEB

FLANGE WIDTH

LENGTH

DEPTH

FLANGE

WEB THICKNESS

M - MISCELLANEOUS BEAM

6 Principal Views and Size Description

Figure 6-16 M-Beam

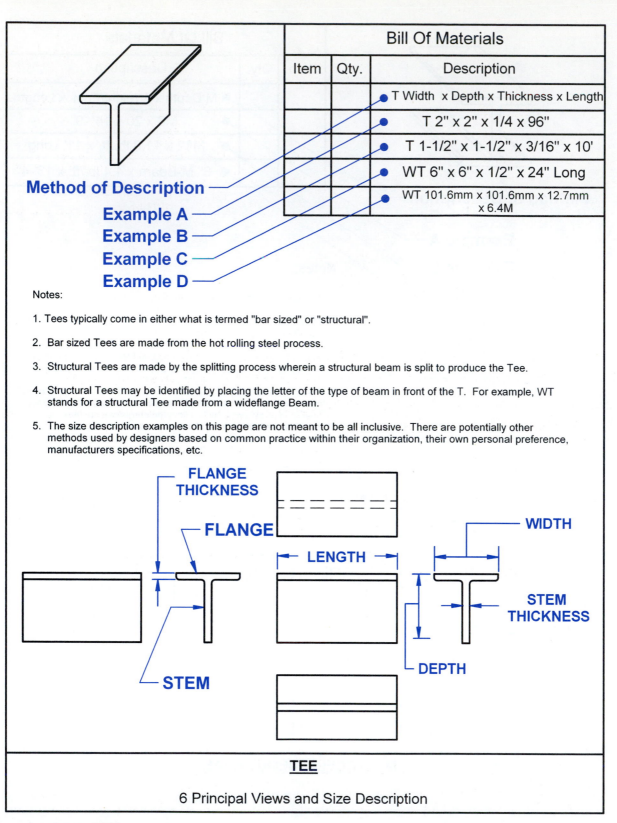

Bill Of Materials

Item	Qty.	Description
		T Width x Depth x Thickness x Length
		T 2" x 2" x 1/4 x 96"
		T 1-1/2" x 1-1/2" x 3/16" x 10'
		WT 6" x 6" x 1/2" x 24" Long
		WT 101.6mm x 101.6mm x 12.7mm x 6.4M

Method of Description

Example A

Example B

Example C

Example D

Notes:

1. Tees typically come in either what is termed "bar sized" or "structural".

2. Bar sized Tees are made from the hot rolling steel process.

3. Structural Tees are made by the splitting process wherein a structural beam is split to produce the Tee.

4. Structural Tees may be identified by placing the letter of the type of beam in front of the T. For example, WT stands for a structural Tee made from a wideflange Beam.

5. The size description examples on this page are not meant to be all inclusive. There are potentially other methods used by designers based on common practice within their organization, their own personal preference, manufacturers specifications, etc.

FLANGE THICKNESS

FLANGE

WIDTH

LENGTH

STEM THICKNESS

DEPTH

STEM

TEE

6 Principal Views and Size Description

Figure 6-17   TEE

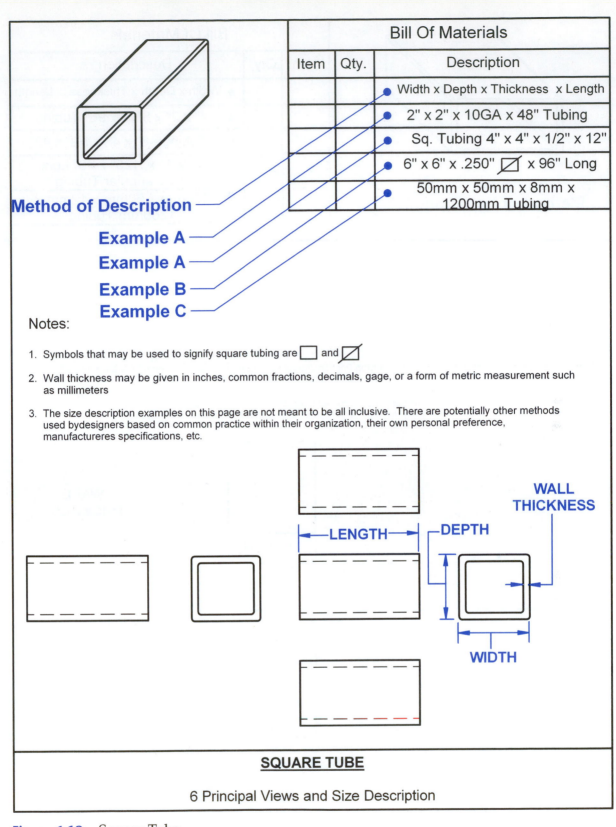

Bill Of Materials		
Item	Qty.	Description
		Width x Depth x Thickness x Length
		2" x 2" x 10GA x 48" Tubing
		Sq. Tubing 4" x 4" x 1/2" x 12"
		6" x 6" x .250" ⊠ x 96" Long
		50mm x 50mm x 8mm x 1200mm Tubing

Method of Description

Example A

Example A

Example B

Example C

Notes:

1. Symbols that may be used to signify square tubing are ☐ and ⊠

2. Wall thickness may be given in inches, common fractions, decimals, gage, or a form of metric measurement such as millimeters

3. The size description examples on this page are not meant to be all inclusive. There are potentially other methods used bydesigners based on common practice within their organization, their own personal preference, manufactureres specifications, etc.

WALL THICKNESS

←LENGTH→ DEPTH

WIDTH

SQUARE TUBE

6 Principal Views and Size Description

Figure 6-18   Square Tube

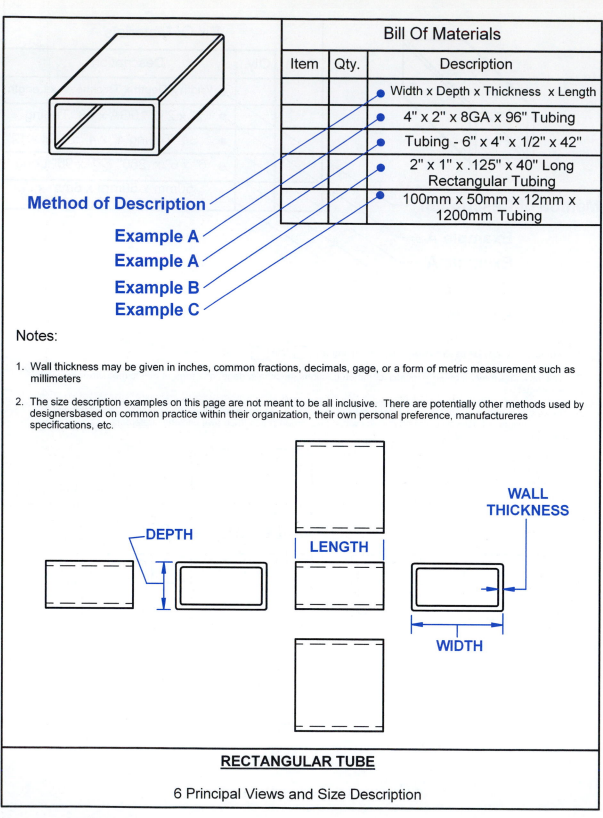

Bill Of Materials

Item	Qty.	Description
		Width x Depth x Thickness x Length
		4" x 2" x 8GA x 96" Tubing
		Tubing - 6" x 4" x 1/2" x 42"
		2" x 1" x .125" x 40" Long Rectangular Tubing
		100mm x 50mm x 12mm x 1200mm Tubing

Method of Description

Example A

Example A

Example B

Example C

Notes:

1. Wall thickness may be given in inches, common fractions, decimals, gage, or a form of metric measurement such as millimeters

2. The size description examples on this page are not meant to be all inclusive. There are potentially other methods used by designersbased on common practice within their organization, their own personal preference, manufactureres specifications, etc.

WALL THICKNESS

DEPTH

LENGTH

WIDTH

RECTANGULAR TUBE

6 Principal Views and Size Description

Figure 6-19 Rectangular Tube

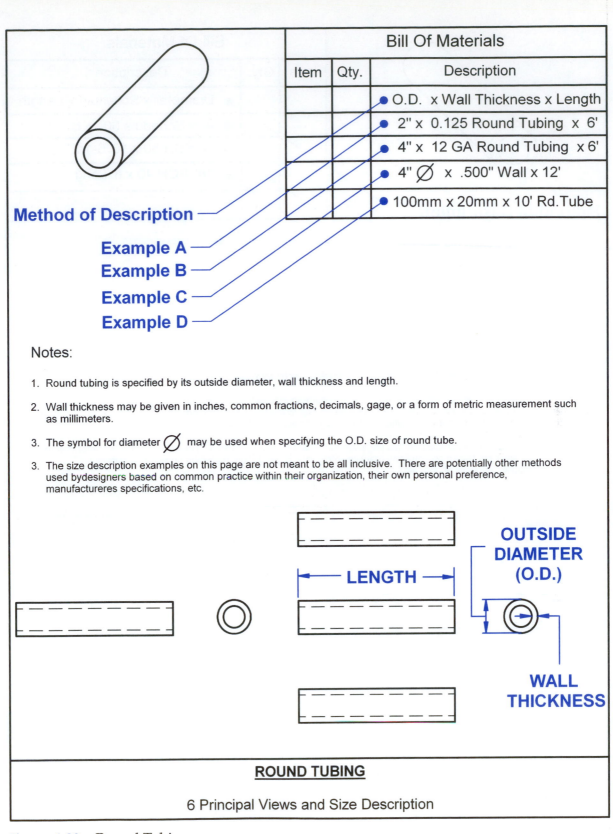

Bill Of Materials		
Item	Qty.	Description
		O.D. x Wall Thickness x Length
		2" x 0.125 Round Tubing x 6'
		4" x 12 GA Round Tubing x 6'
		4" ⌀ x .500" Wall x 12'
		100mm x 20mm x 10' Rd.Tube

Method of Description

Example A

Example B

Example C

Example D

Notes:

1. Round tubing is specified by its outside diameter, wall thickness and length.

2. Wall thickness may be given in inches, common fractions, decimals, gage, or a form of metric measurement such as millimeters.

3. The symbol for diameter ⌀ may be used when specifying the O.D. size of round tube.

3. The size description examples on this page are not meant to be all inclusive. There are potentially other methods used bydesigners based on common practice within their organization, their own personal preference, manufactureres specifications, etc.

LENGTH

OUTSIDE DIAMETER (O.D.)

WALL THICKNESS

ROUND TUBING

6 Principal Views and Size Description

Figure 6-20 Round Tubing

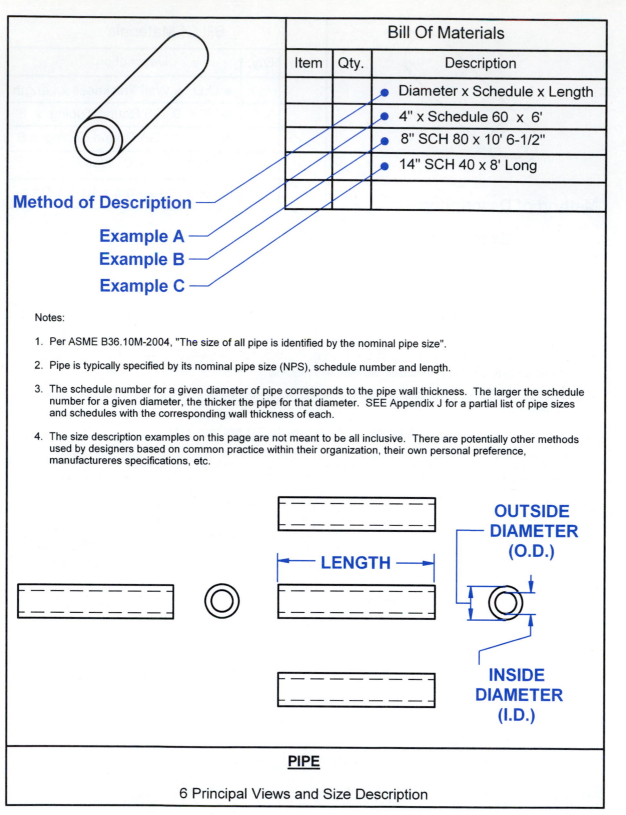

Bill Of Materials		
Item	Qty.	Description
		● Diameter x Schedule x Length
		● 4" x Schedule 60 x 6'
		● 8" SCH 80 x 10' 6-1/2"
		● 14" SCH 40 x 8' Long

Method of Description

Example A

Example B

Example C

Notes:

1. Per ASME B36.10M-2004, "The size of all pipe is identified by the nominal pipe size".

2. Pipe is typically specified by its nominal pipe size (NPS), schedule number and length.

3. The schedule number for a given diameter of pipe corresponds to the pipe wall thickness. The larger the schedule number for a given diameter, the thicker the pipe for that diameter. SEE Appendix J for a partial list of pipe sizes and schedules with the corresponding wall thickness of each.

4. The size description examples on this page are not meant to be all inclusive. There are potentially other methods used by designers based on common practice within their organization, their own personal preference, manufactureres specifications, etc.

LENGTH

OUTSIDE DIAMETER (O.D.)

INSIDE DIAMETER (I.D.)

PIPE

6 Principal Views and Size Description

Figure 6-21 Pipe

CHAPTER 6
Practice Exercise 1

Answer the following questions or define the term.

1. Why is it important to make sure you have the correct metal type for the job?

   _____

2. What steel group does 4140 belong to, based on the AISI-SAE steel classification system?

   _____

3. Based on Table 6.1, what is the nominal carbon content of 1018 steel? _____

4. What does the letter T stand for in the Unified Numbering System (UNS)? _____

5. What do the last five numbers in the UNS system designate? _____

6. Is a piece of steel that measures $\frac{1}{8}''$ × 36″ × 52″ considered sheet or plate? _____

7. How is round bar typically specified? _____

8. Describe the shape and dimensional characteristics of C5 × 9 × 12′. _____

9. How much does a W10 × 88 beam weigh? _____

10. How much does an S20 × 75 beam weigh? _____

11. How is tubing specified? _____

12. How is pipe specified? _____

13. What is the thickness of the legs and the length of the legs for a piece of

    2″ × 2″ × $\frac{1}{4}''$ × 3″ angle? _____

14. What does gauge refer to when discussing sheet metal? _____

15. What do flange and web refer to? _____

   _____

Practice Exercise 2

Match the following right-side (end) views with their corresponding front and top views, which are shown on the following page.

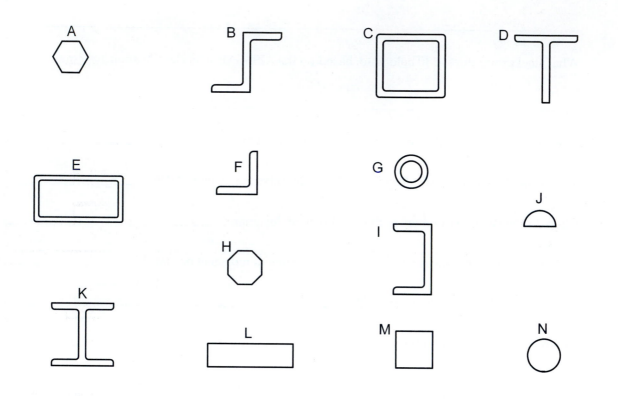

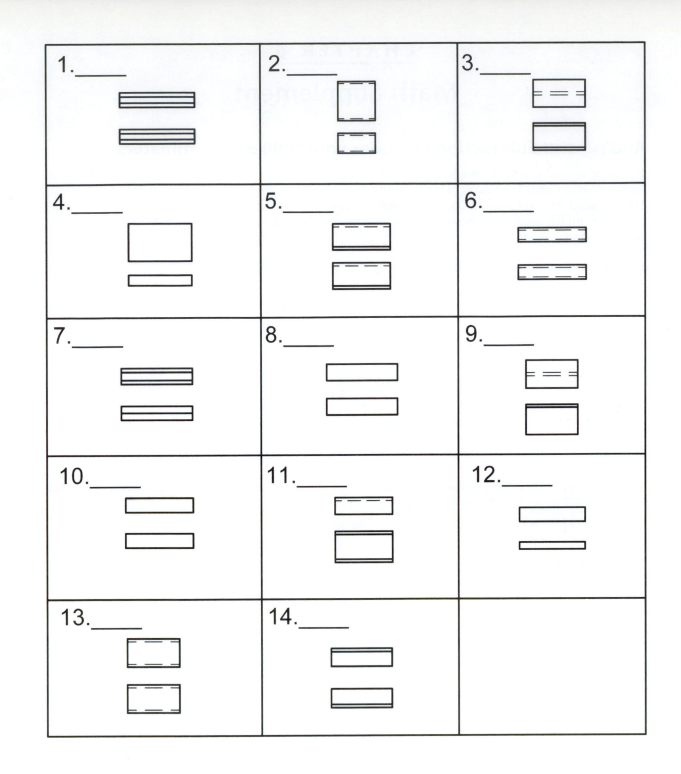

1. _____

2. _____

3. _____

4. _____

5. _____

6. _____

7. _____

8. _____

9. _____

10. _____

11. _____

12. _____

13. _____

14. _____

CHAPTER 6

Math Supplement

Adding and Subtracting Fractions with Unlike Denominators

There are several methods for adding or subtracting fractions that have unlike denominators. We will look at two of them. The first is to find the lowest denominator that is common to all of the fractions. This is called the lowest common denominator (LCD). Once the LCD has been found, follow the procedures from the properties shown in Chapter 5 for adding and subtracting fractions with like denominators.

Example 1: Add the following fractions.

$$\frac{1}{8} + \frac{3}{16} + \frac{3}{4} = ?$$

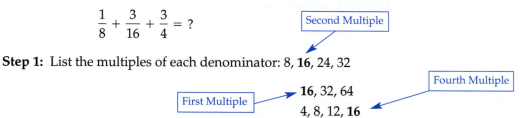

Step 1: List the multiples of each denominator: 8, **16**, 24, 32

16, 32, 64

4, 8, 12, **16**

As you can see, 16 is the smallest number in each list, so it is the LCD.

Step 2: Count the number of multiples required to get to the LCD for each number.

Step 3: Multiply the numerator and the denominator in each fraction by the multiple number for that particular fraction, as found in step 2.

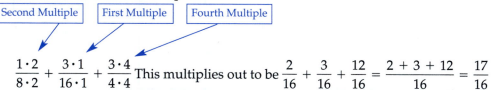

$$\frac{1 \cdot 2}{8 \cdot 2} + \frac{3 \cdot 1}{16 \cdot 1} + \frac{3 \cdot 4}{4 \cdot 4} \text{ This multiplies out to be } \frac{2}{16} + \frac{3}{16} + \frac{12}{16} = \frac{2 + 3 + 12}{16} = \frac{17}{16}$$

The second method for adding fractions that have unlike denominators is to convert the fractions to their decimal equivalent, add the decimal numbers together, and then convert them back to fractions.

Example 2: Add the following fractions.

$$\frac{1}{8} + \frac{3}{16} + \frac{3}{4} = ?$$

Step 1: Change the fractions to their decimal equivalent by dividing the numerator by the denominator. Use longhand division or a calculator, or look up the decimal equivalent on a chart such as the one found in Appendix B at the back of this book.

$$\frac{1}{8} = 8\overline{)1} = .125 \qquad \frac{3}{16} = 16\overline{)3} = .1875 \qquad \frac{3}{4} = 4\overline{)3} = .750$$

Step 2: Add the decimal equivalents.

$$.125 + .1875 + .750 = 1.0625$$

Step 3: If the number is a whole number (1, 2, 3, 4, 5, 6, etc.) you're done. If the number has nonzero numbers following (to the right of) the decimal point, look up those numbers to the right of the decimal point on a decimal to common fraction conversion chart (see Appendix B) and replace the decimal portion with its common fraction equivalent.

In our example, the decimal equivalents add up to 1.0625 which converts to a final answer of $1\frac{1}{16}$.

Math Supplement Practice Exercise 1

Add the following fractions. Write your answer in lowest common denominator form.

1. $\dfrac{1}{8} + \dfrac{3}{8} =$

2. $\dfrac{3}{16} + \dfrac{5}{16} =$

3. $\dfrac{23}{64} + \dfrac{17}{64} =$

4. $\dfrac{3}{4} + \dfrac{1}{4} =$

5. $\dfrac{17}{32} + \dfrac{8}{32} =$

6. $\dfrac{5}{32} + \dfrac{7}{32} =$

7. $\dfrac{1}{8} + \dfrac{3}{16} =$

8. $\dfrac{1}{2} + \dfrac{3}{16} =$

9. $\dfrac{3}{4} + \dfrac{3}{32} =$

10. $\dfrac{1}{2} + \dfrac{1}{4} =$

11. $\dfrac{1}{2} + \dfrac{3}{16} =$

12. $\dfrac{7}{8} + \dfrac{1}{16} =$

Change the following fractions from improper fractions to mixed numbers. Show the fractions in their lowest common denominator form.

13. $\dfrac{5}{4} =$

14. $\dfrac{3}{2} =$

15. $\dfrac{12}{8} =$

16. $\dfrac{66}{64} =$

17. $\dfrac{38}{32} =$

18. $\dfrac{71}{64} =$

Weld Joints

KEY TERMS

Weld joint	Lap joint	T-joint
Butt joint	Edge joint	Edge shape
Corner joint		

OVERVIEW

A weld is made either to join two or more materials together or to add material onto an existing substrate (base material). A **weld joint** is formed where two or more pieces of weldable material are to be joined together by welding. Figure 7-1 shows the five basic classifications of weld joints: the **butt joint**, **corner joint**, **lap joint**, **edge joint**, and **T-joint**. These classifications are based on how the two pieces to be joined are positioned with respect to one another. Beyond the basic classification, there can be many different variations of each basic weld joint. The variations are based on the exact location of each member to be joined and the type of preparation used on each of the members. Material thickness, strength requirements, accessibility for welding, and weld process specifications are all considerations when determining what type of preparation is required.

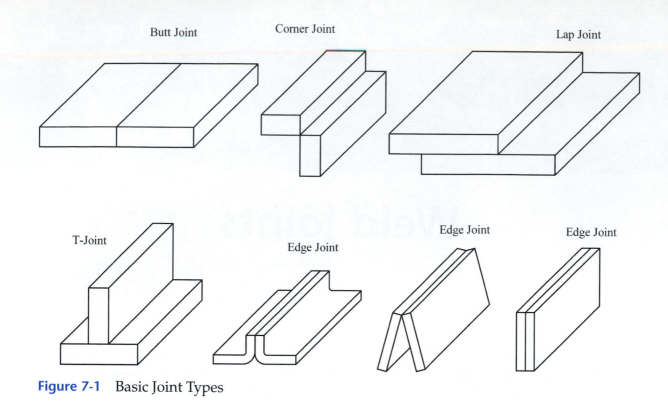

Figure 7-1 Basic Joint Types

EDGE SHAPES

Each of the pieces of base material that make up the particular joint can be prepared to have one of many different types of edge shapes. Figure 7-2 shows the different **edge shape** variations that are typically used. The method used to prepare the edge shapes varies depending on the type of edge shape required, the equipment available, and the material thickness. Many preparation methods are used; a few examples are oxy-fuel, plasma, carbon arc, laser, mechanical beveling equipment, and typical machining equipment such as milling machines and lathes.

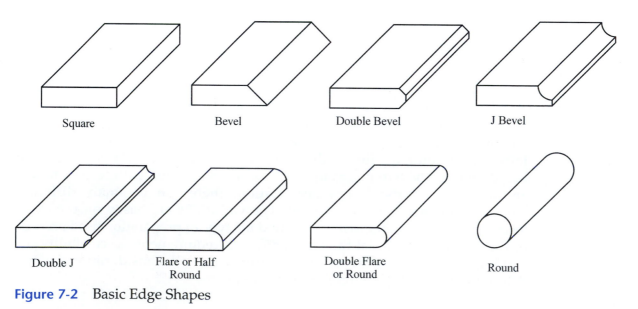

Figure 7-2 Basic Edge Shapes

TYPES OF EDGE PREPARATION FOR BUTT JOINTS

A butt joint occurs where the edges of two pieces to be welded are butted up to each other (see Figure 7-3). A good example of a butt joint is where two 4′ × 8′ sheets of steel are welded together to create one 8′ × 8′ sheet of steel.

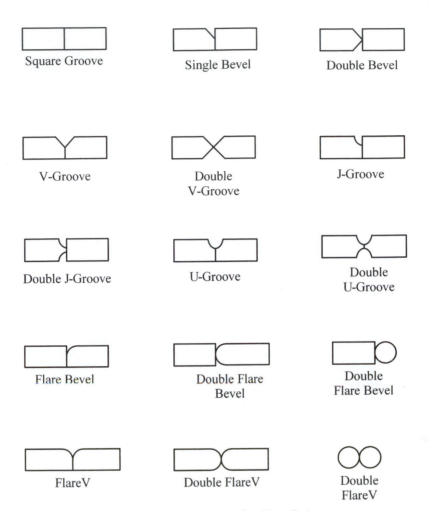

Square Groove   Single Bevel   Double Bevel

V-Groove   Double V-Groove   J-Groove

Double J-Groove   U-Groove   Double U-Groove

Flare Bevel   Double Flare Bevel   Double Flare Bevel

FlareV   Double FlareV   Double FlareV

Figure 7-3   Types of Preparation for Butt Joints

TYPES OF EDGE PREPARATION FOR CORNER JOINTS

A corner joint, shown in Figure 7-4, is formed where two pieces of material come together to form a corner. There are several different variations of the corner joint, with many different types of edge preparation.

TYPES OF PREPARATION FOR LAP JOINTS

A lap joint is formed when two base material surfaces are overlapped together, as shown in Figure 7-5. The edges of the surfaces may be offset or in line with each other.

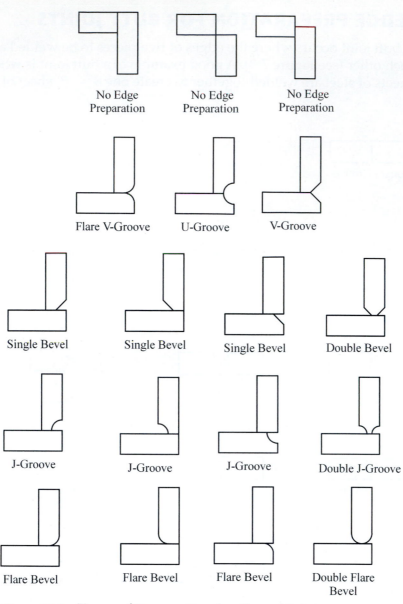

No Edge Preparation	No Edge Preparation	No Edge Preparation

Flare V-Groove	U-Groove	V-Groove

Single Bevel	Single Bevel	Single Bevel	Double Bevel

J-Groove	J-Groove	J-Groove	Double J-Groove

Flare Bevel	Flare Bevel	Flare Bevel	Double Flare Bevel

Figure 7-4 Types of Preparation for Corner Joints

TYPES OF PREPARATION FOR EDGE JOINTS

An edge joint occurs when the surface of one base material is facing the surface of another base material, with the edges of the surfaces aligned closely enough such that the edges can be welded together. Figures 7-6 and 7-7 show different variations of edge joints.

TYPES OF EDGE PREPARATION FOR T-JOINTS

A T-joint occurs where the edge of one base material member faces the surface of another base material member, thus forming a joint that has two sides opposite each other that form inside corners, each measuring approximately 90 degrees between the two base materials. In other words, the joint looks like a T. It can be

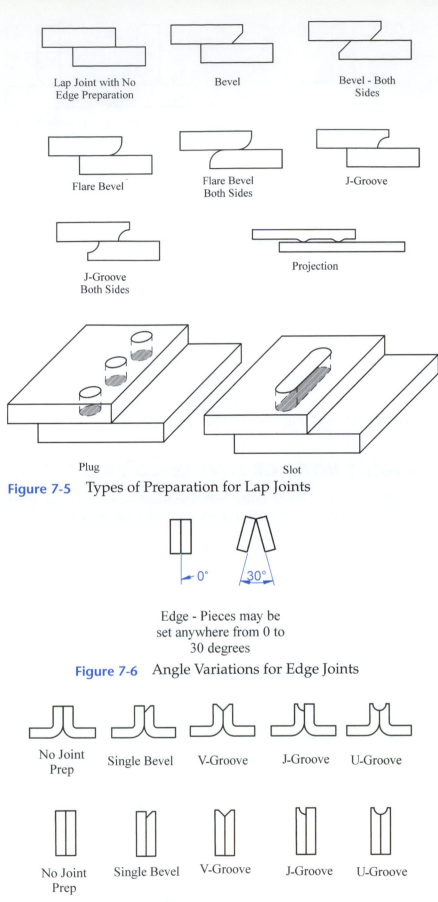

Lap Joint with No Edge Preparation

Bevel

Bevel - Both Sides

Flare Bevel

Flare Bevel Both Sides

J-Groove

J-Groove Both Sides

Projection

Plug

Slot

Figure 7-5 Types of Preparation for Lap Joints

0°

30°

Edge - Pieces may be set anywhere from 0 to 30 degrees

Figure 7-6 Angle Variations for Edge Joints

No Joint Prep

Single Bevel

V-Groove

J-Groove

U-Groove

No Joint Prep

Single Bevel

V-Groove

J-Groove

U-Groove

Figure 7-7 Types of Preparation for Edge Joints

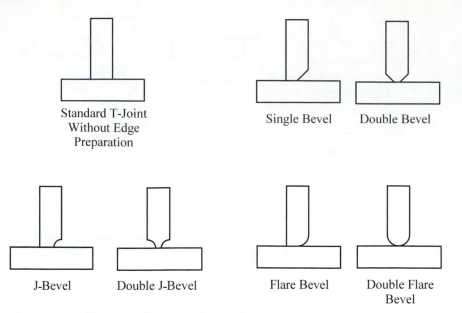

Standard T-Joint
Without Edge
Preparation

Single Bevel   Double Bevel

J-Bevel   Double J-Bevel   Flare Bevel   Double Flare
Bevel

Figure 7-8   Types of Preparation T-Joints

upside down, sideways, or at any angle as long as it forms a T. Figure 7-8 shows the typical types of edge preparation used on T-joints.

PARTS OF A JOINT WITH GROOVE WELD PREPARATION

A. *Root opening:* The distance between the materials to be joined. Other terms used to denote root opening are gap, root gap and joint gap. *Note:* The root opening on a joint may be zero.
B. *Root face:* The unbeveled area of the joint used to control penetration and melt-through. Land is another term used to denote root face. *Note:* The root face on a joint may be zero.
C. *Bevel angle:* The angle at which the bevel is cut. *Note:* The bevel angle on a joint may be zero.
D. *Bevel face:* The surface of the cut bevel.
E. *Groove angle or included angle:*   The total angle of the groove joint (see Figure 7-9). For bevel joints, where only one of the two members is beveled, this equals the bevel angle. For joints where both members are beveled, such as a V-groove joint, this is the total angle of both bevel angles added together. *Note:* The groove angle may be zero.
F. *Compound bevel angle:* The secondary angle cut on a compound bevel.
G. *Groove radius:* The radius of the groove on J-bevel or U-groove weld joints (see Figure 7-10).

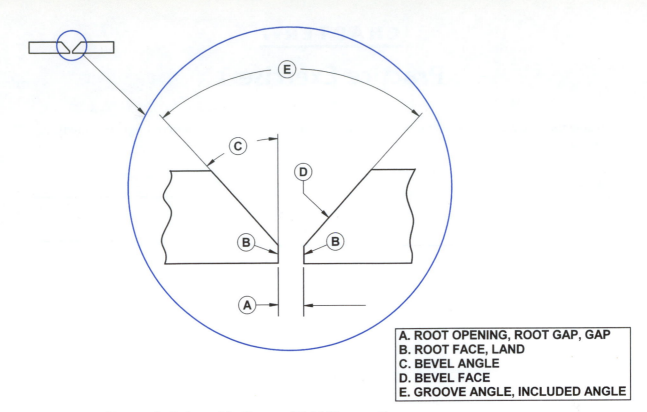

A. ROOT OPENING, ROOT GAP, GAP
B. ROOT FACE, LAND
C. BEVEL ANGLE
D. BEVEL FACE
E. GROOVE ANGLE, INCLUDED ANGLE

Figure 7-9   Parts of a Joint with Groove Weld Preparation

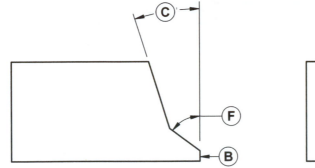
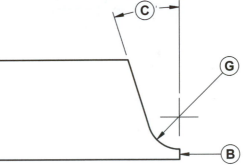

COMPOUND BEVEL   J-BEVEL

B. ROOT FACE, LAND
C. BEVEL ANGLE
F. COMPOUND BEVEL ANGLE
G. GROOVE RADIUS

Figure 7-10   Compound Bevel and J-Bevel Weld Preparation

Practice Exercise 1

1. Name the five basic classifications of weld joints corresponding to the lettered figures shown below.

A. _____

B. _____

C. _____

D. _____

E. _____

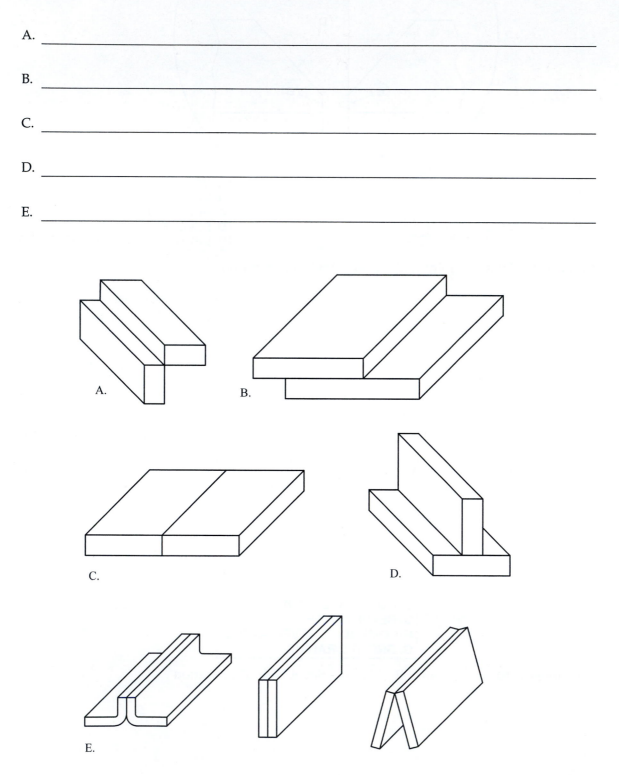

2. Name the eight basic types of edge preparation corresponding to the lettered figures shown below.

A. _____

B. _____

C. _____

D. _____

E. _____

F. _____

G. _____

H. _____

A.

B.

C.

D.

E.

F.

G.

H.

Practice Exercise 2

Sketch end views of the following weld joints in the spaces provided.

Square groove butt joint	
Single bevel butt joint	
V-groove butt joint	
J-groove butt joint	
U-groove butt joint	

Flare bevel butt joint	
Flare-V butt joint	
Single bevel T-joint	
J-bevel T-joint	
Lap joint without edge preparation	

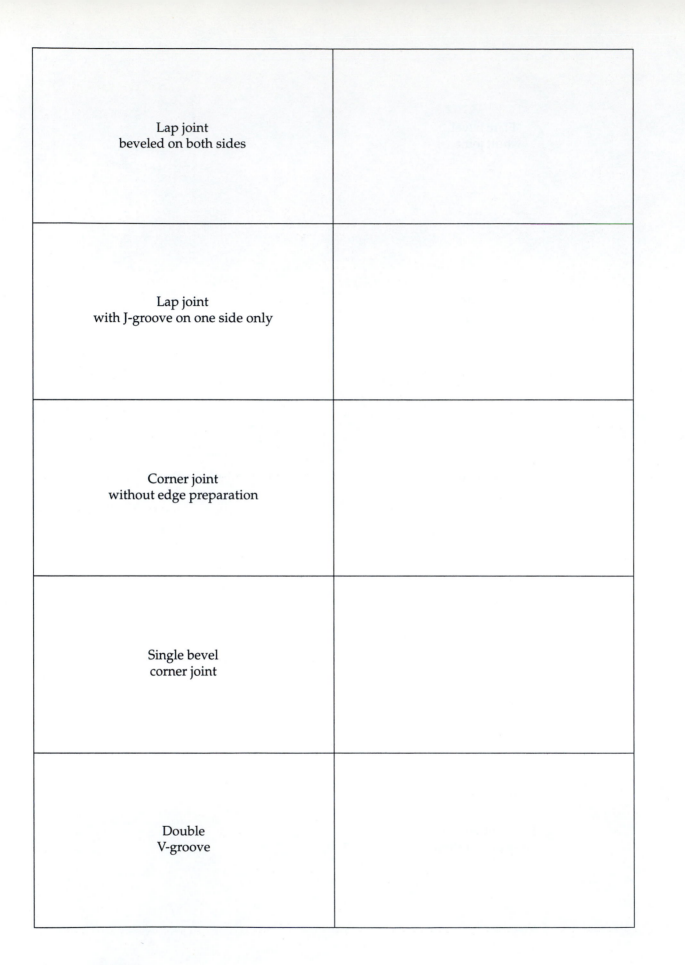

Lap joint
beveled on both sides

Lap joint
with J-groove on one side only

Corner joint
without edge preparation

Single bevel
corner joint

Double
V-groove

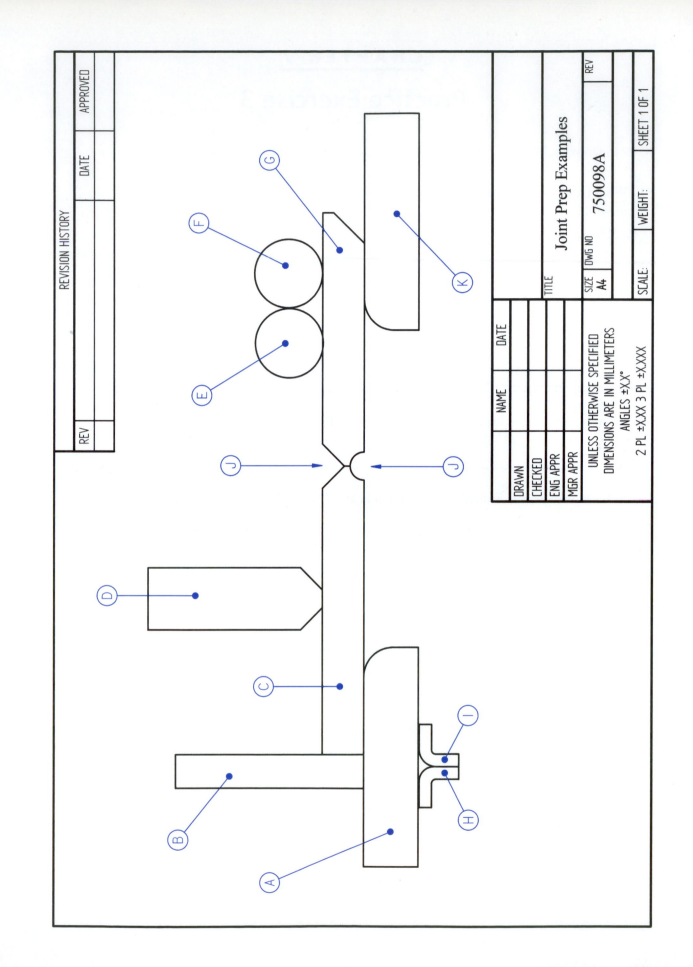

CHAPTER 7
Practice Exercise 3

Refer to Drawing Number 750098A on page 115 for the following questions.

1. What type of joint is formed between parts B and C?

   _____

2. What type of joint is formed between parts C and G?

   _____

3. What type of joint is formed between parts A and B?

   _____

4. What type of joint is formed between parts E and F?

   _____

5. What type of joint is formed between parts H and I?

   _____

6. What type of edge preparation is required on part A?

   _____

7. What types of edge preparation are required on part C?

   _____

8. What type of edge preparation is required on part D?

   _____

9. What types of edge preparation are required on part G?

   _____

10. What types of edge preparation are located at J?

    _____

CHAPTER 7

Math Supplement

Improper Fractions and Mixed Numbers

It is advantageous in the reading of drawings to be able to convert back and forth between improper fractions and mixed numbers. An improper fraction is a fraction that has a numerator that is larger than the denominator. This sometimes happens when fractions are added together or when it is necessary to make mathematical operations easier. 6/4 is an example of an improper fraction; 1⅛ is an example of a mixed number. The following examples show how to convert between the two.

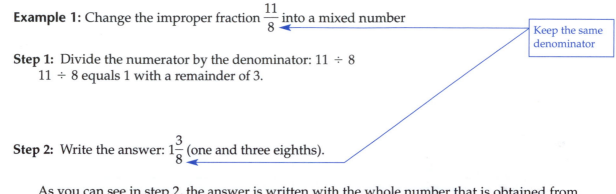

Example 1: Change the improper fraction $\frac{11}{8}$ into a mixed number

Keep the same denominator

Step 1: Divide the numerator by the denominator: 11 ÷ 8
11 ÷ 8 equals 1 with a remainder of 3.

Step 2: Write the answer: $1\frac{3}{8}$ (one and three eighths).

As you can see in step 2, the answer is written with the whole number that is obtained from step 1 and the remainder as the numerator in a fraction where the denominator is the same as it was in the improper fraction you started with.

Example 2: Change the mixed number $3\frac{5}{8}$ into an improper fraction.

Step 1: Multiply the denominator by the whole number: 8 times 3 = 24.

Step 2: Add the numerator to the result of step 1: 24 + 5 = 29.

Step 3: Place the result of step 2 in the numerator of the new fraction: $\frac{29}{}$.

Step 4: Place the original denominator as the denominator in the new fraction: $\frac{29}{8}$.

Answer $= \frac{29}{8}$

Math Supplement Practice Exercise 1

Change the following numbers from improper fractions to mixed numbers.

	Improper Fraction	Mixed Number
1.	$\frac{13}{2}$	
2.	$\frac{19}{16}$	
3.	$\frac{37}{32}$	
4.	$\frac{45}{8}$	
5.	$\frac{37}{32}$	
6.	$\frac{22}{16}$	
7.	$\frac{72}{64}$	

Write the following fractions in their lowest form.

	Mixed Number	Improper Fraction
8.	$4\frac{3}{16}$	
9.	$1\frac{16}{64}$	
10.	$2\frac{24}{32}$	
11.	$8\frac{5}{8}$	
12.	$42\frac{1}{2}$	
13.	$2\frac{13}{32}$	
14.	$4\frac{3}{4}$	
15.	$3\frac{7}{16}$	

Weld Types

OBJECTIVE

- Be able to identify and name the basic weld types and the terms that apply to each

KEY TERMS

Fillet weld	Effective throat	Plug weld
Leg	Groove weld	Slot weld
Toe	Root reinforcement	Spot weld
Face	Face reinforcement	Seam weld
Fillet weld root	Face width and groove width	Projection weld
Joint root	Backing welds	Stud weld
Theoretical throat	Back welds	Surfacing weld
Actual throat	Edge weld	

OVERVIEW

This chapter describes the following weld types: fillet, groove, backing, back, edge, plug, slot, spot, seam, projection, stud, and surfacing.

FILLET WELDS

A **fillet weld** is placed where two surfaces are located perpendicular or approximately perpendicular to each other. They may be used on T-, corner, or lap joints. Figure 8-1 illustrates how each of the different joints has the same orientation of surfaces where the weld is placed. The cross-sectional shape of a fillet weld is approximately triangular in shape.

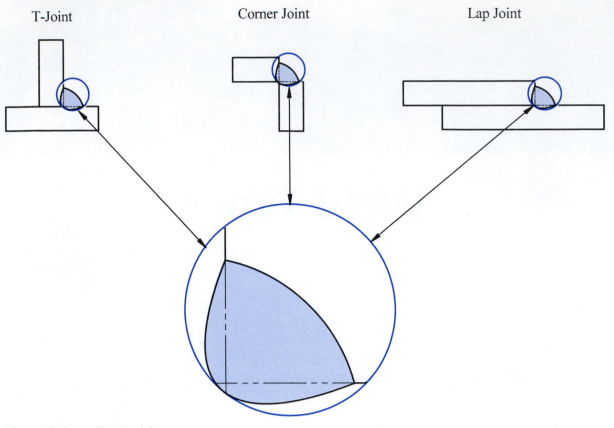

Figure 8-1 Fillet Welds

PARTS OF A FILLET WELD

A. *Leg:* The legs of a fillet weld extend from the intersection of the materials to the toes of the weld.

B. *Toe:* The toes of the weld are where the visual surface of the weld ties into the base metal.

C. *Face:* The front surface of the weld.

D. *Fillet weld root:* Outermost extent of penetration of the weld into the joint where both members are joined.

E. *Joint root:* For fillet welds, the point of intersection of the two members of the joint (see Figure 8-2).

F. *Theoretical throat:* The distance along a perpendicular line from the root of the joint to a theoretical hypotenuse of the largest right triangle that can be formed wholly within the weld.

G. *Actual throat:* Extends from the face of the weld to the root of the weld.

H. *Effective throat:* Extends from the hypotenuse to the root of the weld (see Figure 8-3).

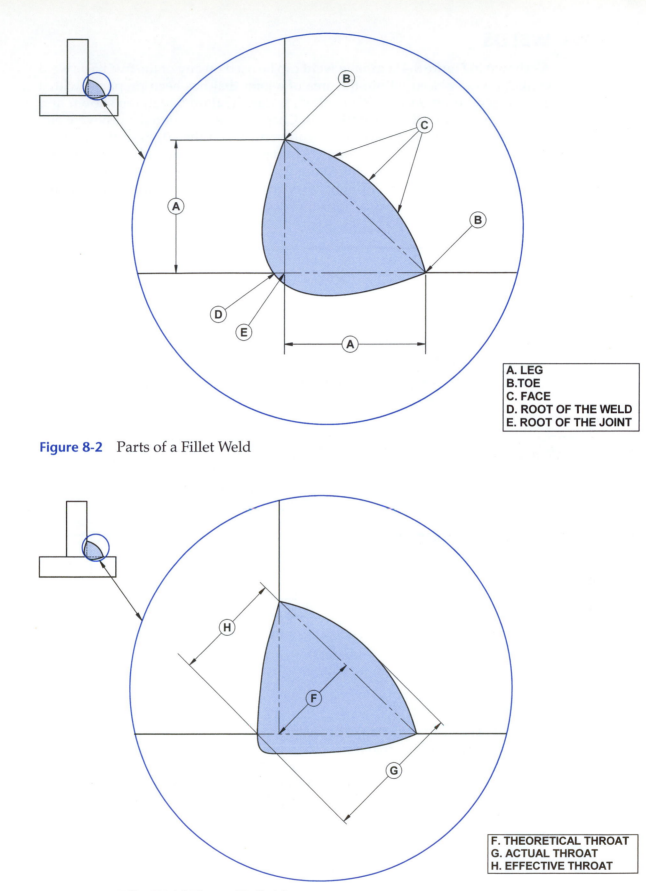

A. LEG
B.TOE
C. FACE
D. ROOT OF THE WELD
E. ROOT OF THE JOINT

Figure 8-2 Parts of a Fillet Weld

F. THEORETICAL THROAT
G. ACTUAL THROAT
H. EFFECTIVE THROAT

Figure 8-3 Fillet Weld Throat Definitions

GROOVE WELDS

As shown in Figure 8-4, a **groove weld** can be used on any of the five basic weld joints. A weld placed within the area of a joint that has been prepared with a groove-type preparation (V, bevel, U, J, flare V, flare bevel), as shown and defined in many of the illustrations in Chapter 7, is considered a groove weld. Other applications of groove welds can occur when the edges of the joint are square, for instance, when the joint is a butt joint or a T-joint with a gap between the two members; a corner joint configured as shown in Figure 8-5; or an edge joint where the two pieces forming the edge are not welded completely over the entire face of the joint, thus creating an **edge weld**.

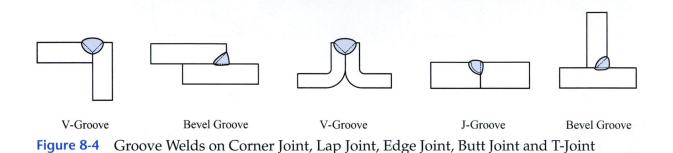

| V-Groove | Bevel Groove | V-Groove | J-Groove | Bevel Groove |

Figure 8-4   Groove Welds on Corner Joint, Lap Joint, Edge Joint, Butt Joint and T-Joint

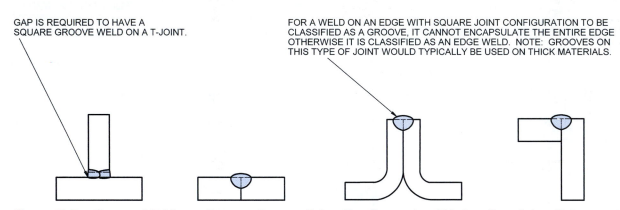

GAP IS REQUIRED TO HAVE A SQUARE GROOVE WELD ON A T-JOINT.

FOR A WELD ON AN EDGE WITH SQUARE JOINT CONFIGURATION TO BE CLASSIFIED AS A GROOVE, IT CANNOT ENCAPSULATE THE ENTIRE EDGE OTHERWISE IT IS CLASSIFIED AS AN EDGE WELD. NOTE: GROOVES ON THIS TYPE OF JOINT WOULD TYPICALLY BE USED ON THICK MATERIALS.

Figure 8-5   Groove Welds on Square Groove Joint Configuration T-Joint, Butt Joint, Edge Joint, and Corner Joint

PARTS OF A GROOVE WELD

A. *Face:* The visible portion of the weld.
B. *Toe:* Where the external face of the weld meets the base metal.
C. *Root:* The deepest point of the weld into the base metal, where the weld metal and both pieces of the base metal are fused together (see Figure 8-6).
D. *Root reinforcement:* The amount of weld bead on the root side that extends beyond the surface of the base metal.
E. *Face reinforcement:* The amount of weld bead on the face side that extends above the surface of the base metal.
F. *Face width:* The width of the weld on the face side.
G. *Groove width:* The distance from one side of the groove joint to the other.
H. *Toes of the weld:* Where the external edges of the weld meet the base metal (see Figure 8-7).

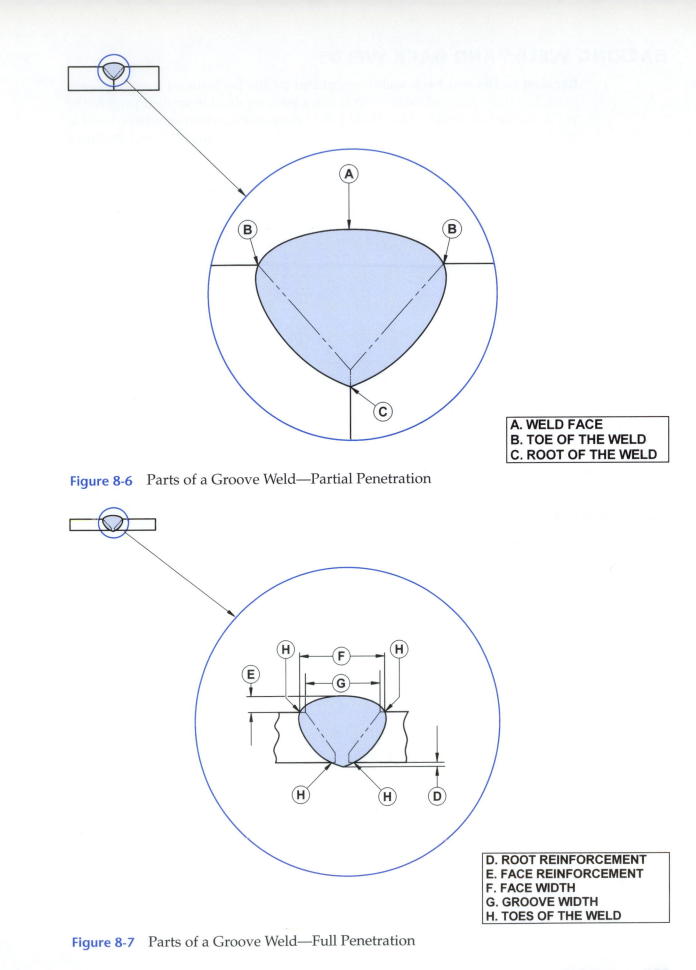

A. WELD FACE
B. TOE OF THE WELD
C. ROOT OF THE WELD

Figure 8-6　Parts of a Groove Weld—Partial Penetration

D. ROOT REINFORCEMENT
E. FACE REINFORCEMENT
F. FACE WIDTH
G. GROOVE WIDTH
H. TOES OF THE WELD

Figure 8-7　Parts of a Groove Weld—Full Penetration

BACKING WELDS AND BACK WELDS

Backing welds and **back welds** are placed on the backside of a groove weld joint. The difference between them is that a backing weld is applied prior to the weld on the front side of the weld joint being made, whereas a back weld is applied after the groove weld has been made. Figure 8-8 shows a backing and a back weld.

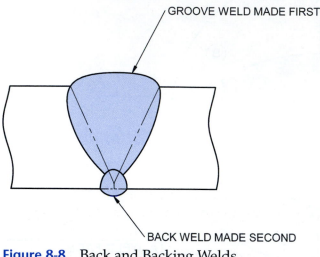
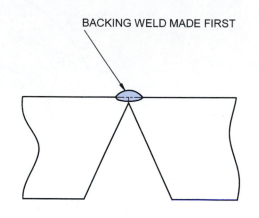

GROOVE WELD MADE FIRST

BACKING WELD MADE FIRST

BACK WELD MADE SECOND

Figure 8-8   Back and Backing Welds

EDGE WELD

An **edge weld** is a weld made on an edge joint. Figure 8-9 shows the placement of an edge weld.

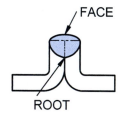
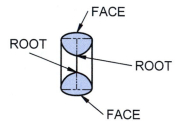

FACE

ROOT

FACE

ROOT

ROOT

FACE

Figure 8-9   Edge Weld

PLUG WELD

A **plug weld** partially or completely fills a circular hole in a weld joint, as shown in Figure 8-10. Plug welds may be left as welded, or they may be finished so that the welded area is nearly unseen.

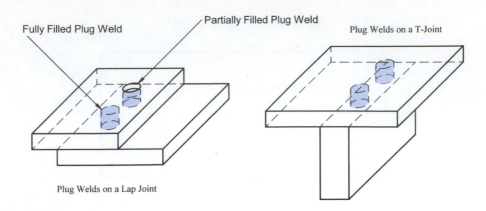

Figure 8-10 Plug Welds

SLOT WELD

A **slot weld** partially or completely fills a slotted hole in a weld, as shown in Figure 8-11. Slot welds may be left as welded, or they may be finished so that the welded area is nearly unseen.

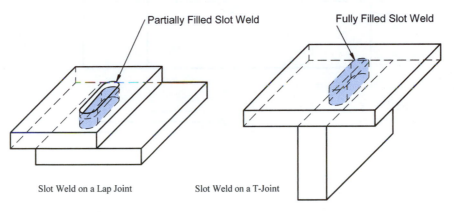

Figure 8-11 Slot Welds

SPOT WELD

A **spot weld** is a solid circular (spot) weld made at the intersection between two faying surfaces. Spot welds can be made from a resistance-type process or a process with enough power density to penetrate through one surface down into the other, as shown in Figure 8-12.

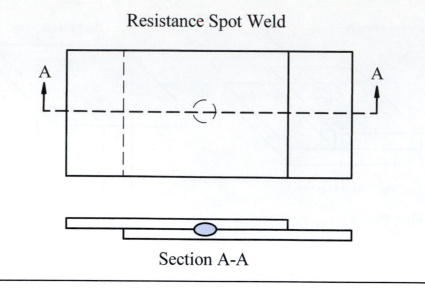

Resistance Spot Weld

Section A-A

Spot Weld from One Sided Process

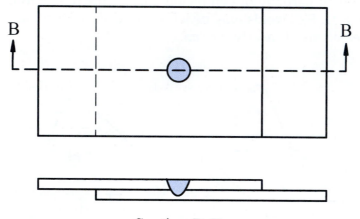

Section B-B

Figure 8-12   Spot Welds

SEAM WELD

A **seam weld** is a longitudinal weld made at the intersection between two faying surfaces. Seam welds can be made from a resistance-type process or a process with enough power density to penetrate through one surface down into the other, as shown in Figure 8-13.

PROJECTION WELD

A **projection weld** is made by the projection welding process, whereby welds are made at projections in the base material between two overlapping surfaces. The projections provide a predetermined spot with a specific size for current coming from a resistance-type welding machine to pass through (see Figure 8-14).

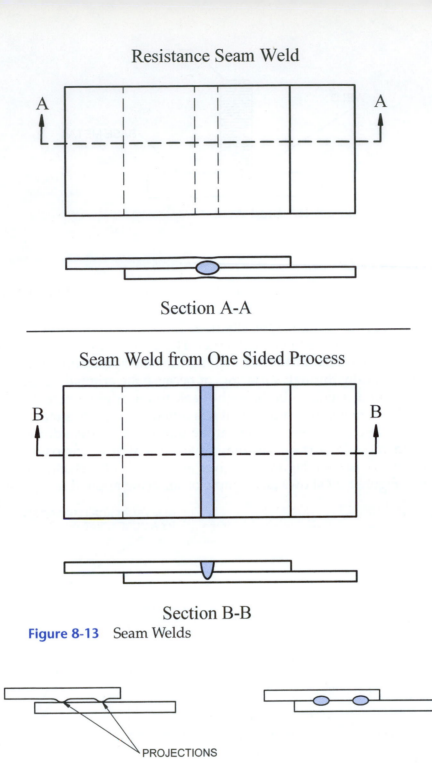

Resistance Seam Weld

Section A-A

Seam Weld from One Sided Process

Section B-B

Figure 8-13   Seam Welds

PROJECTIONS

Figure 8-14   Projection Welds

STUD WELD

A **stud weld** is a weld that attaches a stud to a surface. There are many different kinds of studs, and they come with varying sizes and specifications. Studs may be fully threaded, partially threaded, or not threaded (see Figure 8-15).

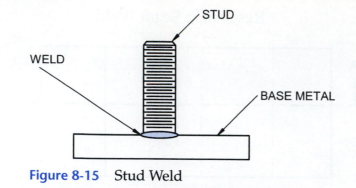

Figure 8-15   Stud Weld

SURFACING WELD

A **surfacing weld** is used to build up surfaces with weld metal. The process of surfacing is also referred to as cladding, overlay, overlaying, pad welding, or padding. Surfacing is typically used for two main reasons. First, it can simply build up a surface to make it larger. This is typically done because a portion of the base material is not large enough for the application and can be enlarged economically through surfacing, or because the original surface has been worn down and needs to be brought back to its original size. Second, surfacing applies a material to a base that has different properties than the base. Some of the different properties that can be improved by the addition of a surfacing weld include surface hardness, wear resistance, corrosion resistance, and resistance to heat. Figure 8-16 gives an example of surfacing that is applied to a bar, Figure 8-17 shows an example of surfacing applied to a plate.

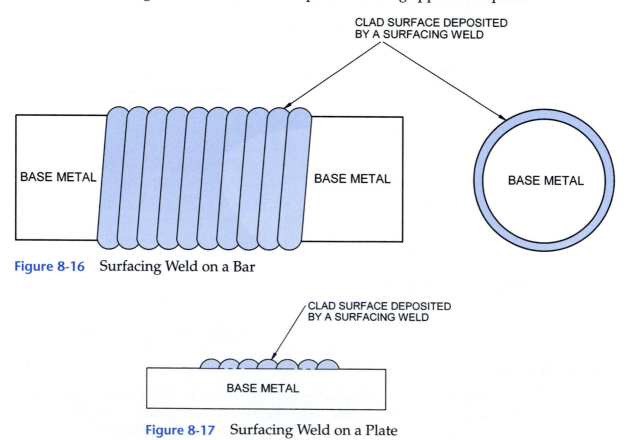

Figure 8-16   Surfacing Weld on a Bar

Figure 8-17   Surfacing Weld on a Plate

Practice Exercise 1

Answer the following questions.

1. What joints can fillet welds be used on? _____

2. What is the approximate cross-sectional shape of a fillet weld? _____

3. What is the face of a weld? _____

4. What joints can a groove weld be used on? _____

5. What is root reinforcement? _____

6. Is a backing weld made before or after a groove weld? _____

7. Would a plug or a slot weld be used on a butt joint? _____

8. How are spot and seam welds similar? _____

   _____

   _____

9. What is a stud weld? _____

10. What other names are used for surfacing welds? _____

11. Why would a part need a clad surface? _____

12. Name some of the properties that can be improved by the surfacing process. _____

   _____

   _____

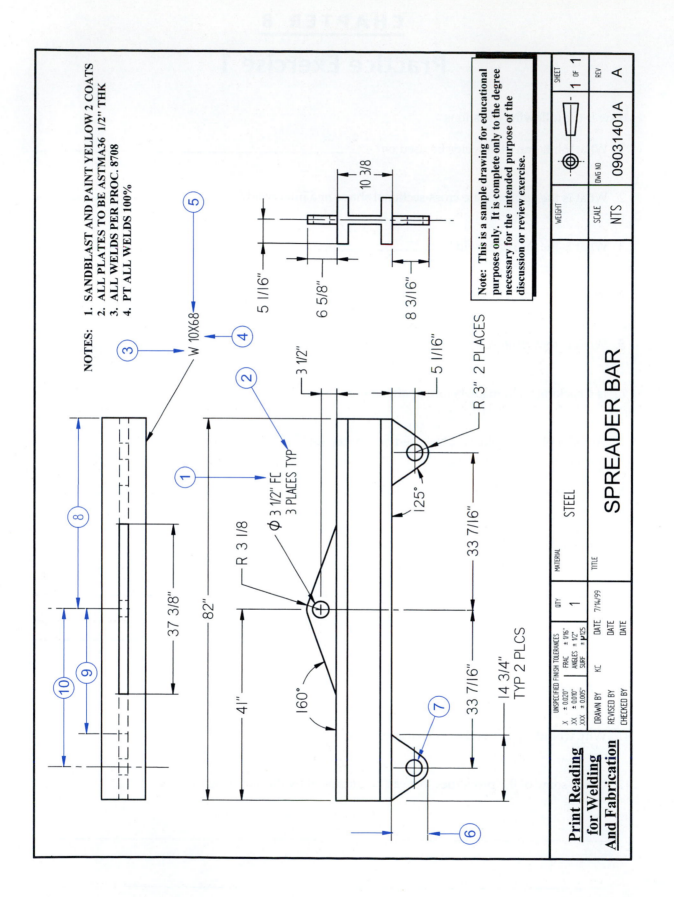

NOTES:

1. SANDBLAST AND PAINT YELLOW 2 COATS
2. ALL PLATES TO BE ASTMA36 1/2" THK
3. ALL WELDS PER PROC. 8708
4. PT ALL WELDS 100%

W 10X68

Ø 3 1/2" FC
3 PLACES TYP

R 3 1/8

R 3" 2 PLACES

125°

160°

82"

41"

37 3/8"

33 7/16"

33 7/16"

14 3/4"
TYP 2 PLCS

3 1/2"

5 1/16"

5 1/16"

6 5/8"

10 3/8

8 3/16"

Note: This is a sample drawing for educational purposes only. It is complete only to the degree necessary for the intended purpose of the discussion or review exercise.

MATERIAL		STEEL
TITLE		SPREADER BAR

QTY	1

UNSPECIFIED FINISH TOLERANCES		
X ± 0.020"	FRAC ± 1/16"	
XX ± 0.010"	ANGLES ± 1/2°	
XXX ± 0.005"	SURF ∜125	

DRAWN BY	KC	DATE 7/14/99
REVISED BY		DATE
CHECKED BY		DATE

**Print Reading
for Welding
And Fabrication**

WEIGHT

SCALE NTS

DWG NO 09031401A

SHEET 1 OF 1

REV A

CHAPTER 8

Practice Exercise 2

Refer to Drawing Number 09031401A Rev. A on page 130 for the following questions.

1. What does FC mean? _____

2. What does TYP mean? _____

3. What does W stand for? _____

4. What does the 10 in W10X68 mean? _____

5. What does the 68 in W10X68 mean? _____

6. What distance is indicated by blue balloon 6? _____

7. What is the diameter of the hole shown at blue balloon 7? _____

8. What distance is indicated by blue balloon 8? _____

9. What distance is indicated by blue balloon 9? _____

10. What distance is indicated by blue balloon 10? _____

11. What is the maximum overall length that the spreader bar can be made? _____

12. What material is the spreader bar made from? _____

13. What views are shown? _____

14. How are the welds to be made? _____

15. What color is the spreader bar to be painted? _____

CHAPTER 8

Math Supplement

Addition and Subtraction with Mixed Numbers

Addition and subtraction operations involving mixed numbers can be accomplished through many different methods. One method is to use a calculator or a decimal chart to convert the fractional portions of the mixed numbers to decimals so that the problem becomes simple addition or subtraction. Several other methods of working these types of problems without a calculator or decimal chart are shown below.

Addition of a whole number and a mixed number

When adding a whole number and a mixed number, add the two whole numbers together and carry the fraction.

Example 1: $7\dfrac{5}{8} + 3 = 10\dfrac{5}{8}$

Subtraction of a whole number from a mixed number

When subtracting a whole number from a mixed number, subtract the whole number from the whole number and carry the fraction.

Example 2: $7\dfrac{5}{8} - 3 = 4\dfrac{5}{8}$

Addition of two mixed numbers The following steps can be used for adding a mixed number to another mixed number.

Step 1: Add all the fractional numbers together.
Step 2: If the resulting fraction is improper, convert it to a mixed number.
Step 3: Add all the whole numbers together.
Step 4: Add the resultant whole number to the fractional portion of the mixed number obtained in step 2 and write the answer.

Example 3: Add the following:

$$7\dfrac{5}{8} + 4\dfrac{13}{16}$$

Step 1: $\dfrac{5}{8} + \dfrac{13}{16} = \dfrac{23}{16}$

Step 2: $\dfrac{23}{16} = 1\dfrac{7}{16}$

Step 3: $7 + 4 + 1 = 12$

Step 4: $12 + \dfrac{7}{16} = 12\dfrac{7}{16}$

Subtraction of two mixed numbers

The following steps can be used for subtracting a mixed number from another mixed number.

Step 1: Perform any borrowing operations as required and rewrite the problem.
Step 2: Perform the required subtraction on the fractional portion of the mixed numbers.
Step 3: Perform the required subtraction operation on all the whole numbers.
Step 4: Add the resultant whole number to the fractional portion of the mixed number obtained in step 2 and write the answer.

Example 3: Subtract the following:

$$7\frac{5}{8} - 4\frac{7}{8}$$

Step 1: Borrow 1 unit, which is $\frac{8}{8}$ from $7\frac{5}{8}$ and rewrite the problem as follows:

$$6\frac{13}{8} - 4\frac{7}{8}$$

Step 2: $\dfrac{13}{8} - \dfrac{7}{8} = \dfrac{6}{8} = \dfrac{3}{4}$

Step 3: $6 - 4 = 2$

Step 4: $2 + \dfrac{3}{4} = 2\dfrac{3}{4}$

Subtraction of a mixed number from a whole number

The following steps can be used for subtracting a mixed number from a whole number.

Step 1: Borrow 1 unit from the whole number and rewrite the problem as the subtraction of a mixed number from another mixed number.
Step 2: Perform the required subtraction on the fractional portion of the mixed numbers.
Step 3: Perform the required subtraction operation on all the whole numbers.
Step 4: Add the resultant whole number to the fractional portion of the mixed number obtained in step 2 and write the answer.

Example 3: Subtract the following:

$$27 - 16\frac{17}{32}$$

Step 1: Borrow 1 unit, which is $\frac{32}{32}$ from 27 and rewrite the problem as follows:

$$26\frac{32}{32} - 16\frac{17}{32}$$

Step 2: $\dfrac{32}{32} - \dfrac{17}{32} = \dfrac{15}{32}$

Step 3: $26 - 16 = 10$

Step 4: $10 + \dfrac{15}{32} = 10\dfrac{15}{32}$

Math Supplement Practice Exercise 1

Answer the following. Write your answer in lowest common denominator form.

1. $7\frac{3}{8} - 5 =$

2. $3\frac{7}{16} + \frac{5}{16} =$

3. $3\frac{23}{64} - 1\frac{17}{64} =$

4. $11\frac{3}{4} - 4\frac{1}{4} =$

5. $6\frac{17}{32} + 4\frac{8}{32} =$

6. $33\frac{13}{32} - 1\frac{7}{32} =$

7. $5\frac{3}{8} - \frac{3}{16} =$

8. $8 - 4\frac{3}{16} =$

9. $6\frac{3}{4} - 1\frac{3}{32} =$

10. $5\frac{1}{2} - \frac{1}{4} =$

11. $1'7\frac{1}{2} - 3\frac{7}{32} =$

12. $3'11\frac{7}{8} - 1'2\frac{1}{16} =$

13. $6\frac{3}{4} - \frac{1}{8} =$

14. $7'0'' - \frac{1}{16} =$

15. $1\frac{3}{8} - \frac{5}{16} =$

16. $48\frac{63}{64} - 6 =$

17. $4'0\frac{28}{32} - \frac{3}{16} =$

18. $12 - \frac{17}{32} =$

Introduction to Welding Symbols

OBJECTIVES

- Know what welding symbols are and why they are used
- Know the standard weld symbols
- Be able to interpret basic welding symbols properly

KEY TERMS

Welding symbol	Weld symbols	Both sides
Arrow	Arrow side	Combination weld symbols
Reference line	Other side	Multiple reference lines
Tail		

OVERVIEW

When a design requires welds to be made, the drawings must provide information about the location, size, and type of welds required. The American Welding Society (AWS) has developed a standard system for representing this information. The system is based on what is called a **welding symbol**. This chapter covers the basics of the AWS system. The complete AWS welding symbol will be covered in greater detail in Chapter 10. For now, let's begin learning about the welding symbol by looking at its basic parts, the **arrow**, the **reference line**, and (when needed) the **tail** (see Figure 9-1).

The arrow is used to point to the weld joint where the weld will be made. The reference line is drawn horizontally, and the information that shows the type of weld required is placed under or over it. The tail provides an area for

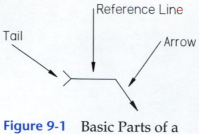

Figure 9-1 Basic Parts of a Welding Symbol

reference information when needed. The arrow may point in any direction (as shown in Figure 9-2) as long as it is pointing at the weld joint.

Figure 9-2 Correctly Drawn Welding Symbols

Welding symbols drawn with their reference lines at angles or without both a reference line and an arrow are drawn incorrectly. See Figure 9-3.

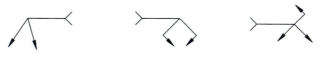

Figure 9-3 Incorrectly Drawn Welding Symbols

More than one arrow may be attached to the end of the reference line in order to point to more than area. See Figure 9-4.

Figure 9-4 Multiple Arrow Symbols

ARROW SIDE AND OTHER SIDE OF WELD JOINTS

Weld joints are designated with an **arrow side** and an **other side**. The arrow side of the joint is the side that the arrow of the welding symbol points to, while the other side is the side opposite the welding symbol arrow. These designations are shown in Figure 9-5.

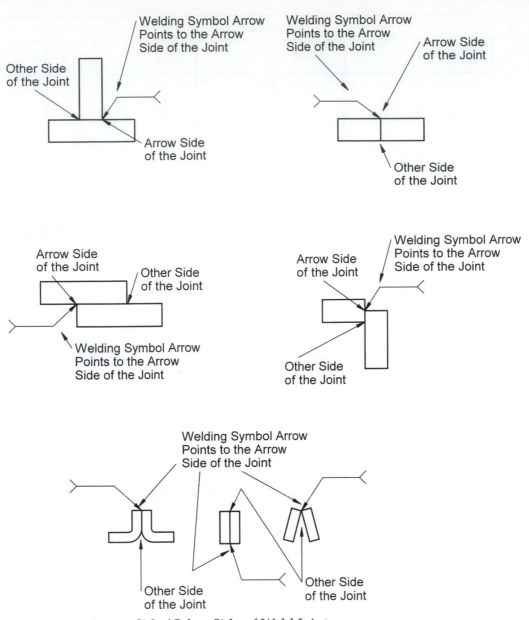

Figure 9-5   Arrow Side/Other Side of Weld Joints

WELD SYMBOLS

Weld symbols are shown in Figure 9-6. They describe the type of weld required at the weld joint. They also describe the type of joint preparation, if required. As you can see when reviewing the weld symbols, the shapes of many of them visually match the joints described in Chapter 7.

The symbols are placed on the reference line of the welding symbol to designate the type of weld required. If the weld is to be made on the arrow side of the weld joint, the weld symbol is placed on the lower side (arrow side) of the reference line. If the weld is to be made on the other side of the weld joint, the

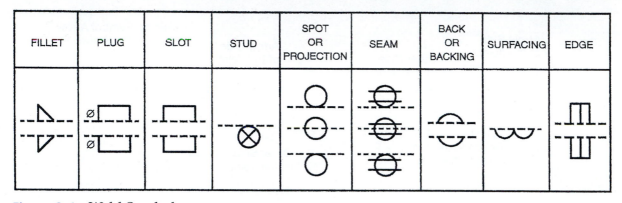

GROOVE							
SQUARE	SCARF	V	BEVEL	U	J	FLARE-V	FLARE-BEVEL

FILLET	PLUG	SLOT	STUD	SPOT OR PROJECTION	SEAM	BACK OR BACKING	SURFACING	EDGE

Figure 9-6 Weld Symbols *(AWS Weld Symbols AWS A2.4:2007, Figure 1 reproduced and adapted with permission from the American Welding Society (AWS), Miami, FL.)*

weld symbol is placed above the reference line. When welds are to be made on **both sides** of the weld joint, weld symbols are placed on both sides of the reference line. Figure 9-7 shows examples of each of these designations. Spot and seam welds can straddle the reference line when it makes no difference which side the weld is made from.

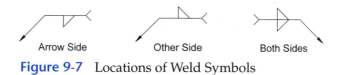

| Arrow Side | Other Side | Both Sides |

Figure 9-7 Locations of Weld Symbols

Now that we know what welding symbols are and how they are to be used, let's look at the welds and weld placements for each of the weld symbols. Figures 9-8 through 9-19 show each of the AWS weld symbols that were shown in Figure 9-6, with an applicable weld joint, weld, and arrow or other side designation.

FILLET WELDS

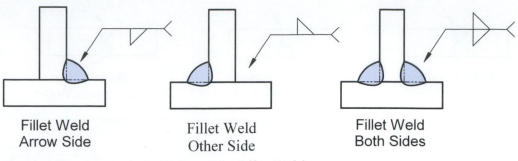

Fillet Weld
Arrow Side

Fillet Weld
Other Side

Fillet Weld
Both Sides

Figure 9-8   Arrow Side/Other Side Fillet Welds

GROOVE WELDS

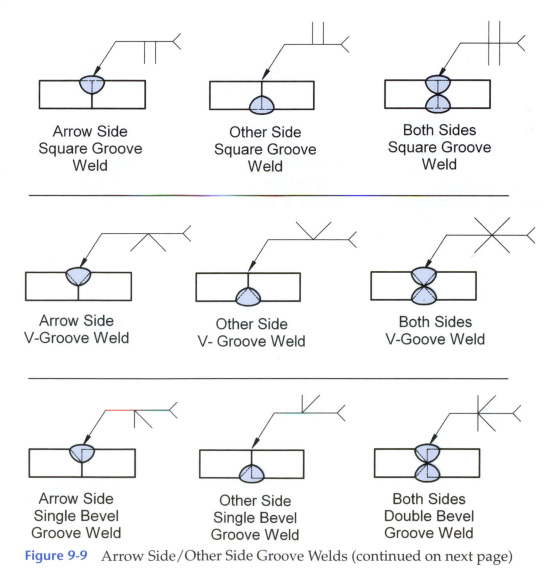

Arrow Side
Square Groove
Weld

Other Side
Square Groove
Weld

Both Sides
Square Groove
Weld

Arrow Side
V-Groove Weld

Other Side
V- Groove Weld

Both Sides
V-Goove Weld

Arrow Side
Single Bevel
Groove Weld

Other Side
Single Bevel
Groove Weld

Both Sides
Double Bevel
Groove Weld

Figure 9-9   Arrow Side/Other Side Groove Welds (continued on next page)

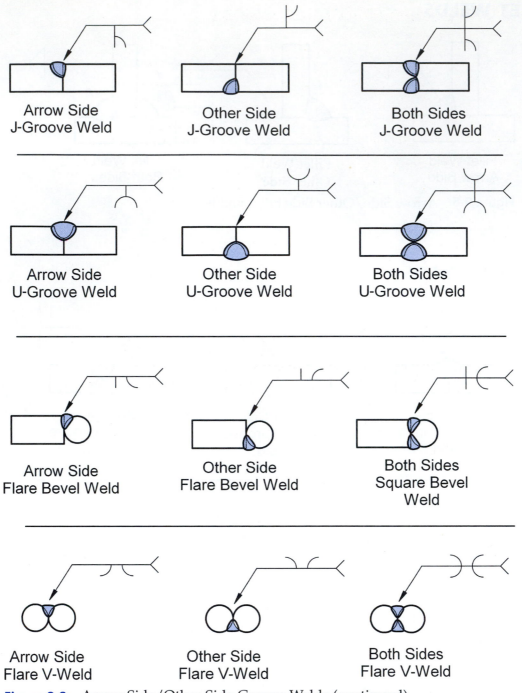

Arrow Side
J-Groove Weld

Other Side
J-Groove Weld

Both Sides
J-Groove Weld

Arrow Side
U-Groove Weld

Other Side
U-Groove Weld

Both Sides
U-Groove Weld

Arrow Side
Flare Bevel Weld

Other Side
Flare Bevel Weld

Both Sides
Square Bevel
Weld

Arrow Side
Flare V-Weld

Other Side
Flare V-Weld

Both Sides
Flare V-Weld

Figure 9-9 Arrow Side/Other Side Groove Welds (continued)

BACK WELDS AND BACKING WELDS

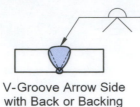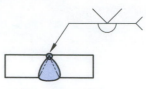

V-Groove Arrow Side
with Back or Backing
Weld on Other Side

V-Groove Other Side
with Back or Backing
Weld on Arrow Side

Figure 9-10   Arrow Side/Other Side Back & Backing Welds

EDGE WELDS

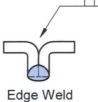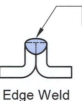

Edge Weld
Arrow Side

Edge Weld
Other Side

Edge Weld
Both Sides

Figure 9-11   Arrow Side/Other Side Edge Welds

SURFACING WELDS

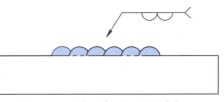

Figure 9-12   Surfacing Welds

SPOT AND PROJECTION WELDS

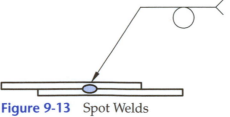

Figure 9-13   Spot Welds

SEAM WELDS

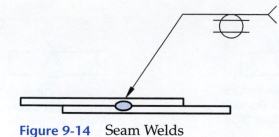

Figure 9-14   Seam Welds

STUD WELDS

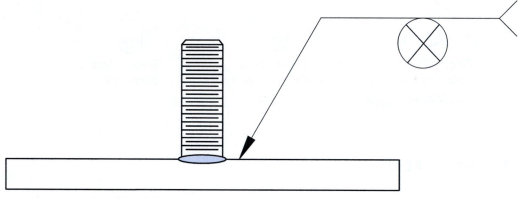

Figure 9-15   Stud Weld

COMBINATION WELD SYMBOLS

Weld joints that require more than one type of weld can be shown by the use of a **combination weld symbol**. The weld performed first is the weld attached to the reference line. See Figures 9-16 and 9-17.

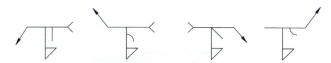

Figure 9-16   Combination Weld Symbols

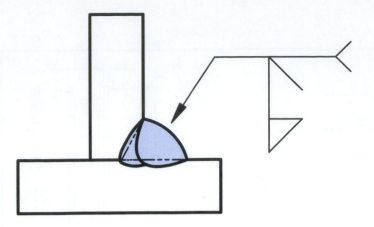

Single Bevel Arrow Side followed by a Fillet Arrow Side

Figure 9-17 Example of Combination Weld and Weld Symbol

MULTIPLE REFERENCE LINES

Another method for showing combination welds is by the use of **multiple reference lines**. When multiple reference lines are used, the order in which the operations are to be performed begins with the reference line that is closest to the arrow. The second set of operations to be completed is shown on the line second closest to the arrow, and so on, as shown in Figures 9-18 and 9-19.

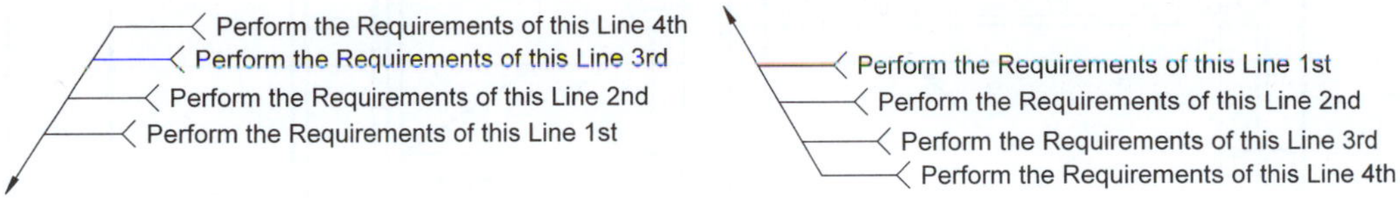

Perform the Requirements of this Line 4th
Perform the Requirements of this Line 3rd
Perform the Requirements of this Line 2nd
Perform the Requirements of this Line 1st

Perform the Requirements of this Line 1st
Perform the Requirements of this Line 2nd
Perform the Requirements of this Line 3rd
Perform the Requirements of this Line 4th

Figure 9-18 Multiple Reference Lines

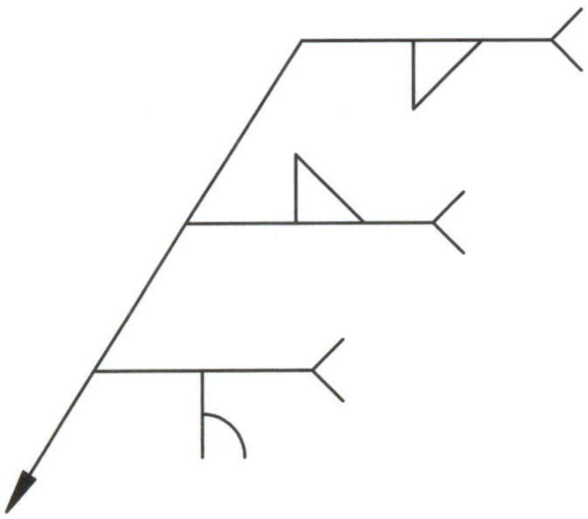

1st Operation: J-Groove on the Arrow Side of the Joint
2nd Operation: Fillet Weld on Other Side of the Joint
3rd Operation: Fillet Weld on the Arrow Side of the Joint

Figure 9-19 Example of Multiple Reference Line Welding Symbol

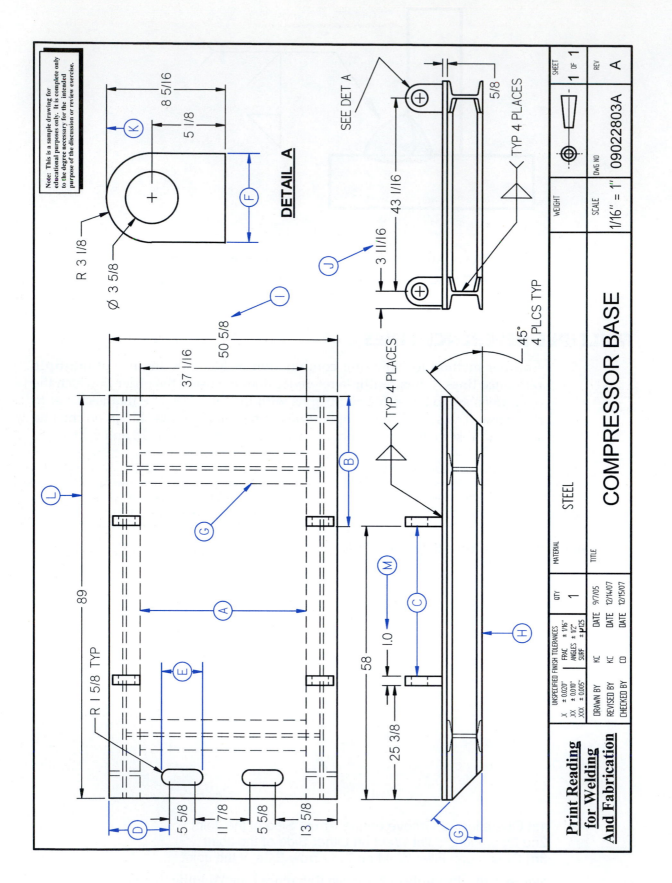

DETAIL A

Note: This is a sample drawing for educational purposes only. It is complete only to the degree necessary for the intended purpose of the discussion or review exercise.

SEE DET A

TYP 4 PLACES

45°
4 PLCS TYP

TYP 4 PLACES

R 1 5/8 TYP

UNSPECIFIED FINISH TOLERANCES		MATERIAL	QTY
X ± 0.020"	FRAC ± 1/16"	STEEL	1
XX ± 0.010"	ANGLES ± 1/2"		
XXX ± 0.005"	SURF ∇ 125		

DRAWN BY	KC	DATE	9/7/05
REVISED BY	KC	DATE	12/14/07
CHECKED BY	CD	DATE	12/15/07

TITLE

COMPRESSOR BASE

| WEIGHT | |
| SCALE | 1/16" = 1" |

| DWG NO | 09022803A |
| SHEET 1 OF 1 | REV A |

Print Reading for Welding And Fabrication

CHAPTER 9

Practice Exercise 1

Refer to Drawing Number 09022803A REV A on page 144 and answer the following questions.

1. What is the tolerance that applies to dimension M? _____

2. What does line H represent? _____

3. What is the angle at G? _____

4. What is the minimum condition of dimension I? _____

5. What is the maximum condition of dimension I? _____

6. What is the minimum condition of dimension J? _____

7. What is the maximum condition of dimension J? _____

8. What type of line is line K? _____

9. What is distance A? _____

10. What is distance B? _____

11. What is distance C? _____

12. What is distance D? _____

13. What is distance E? _____

14. What is distance F? _____

15. What does line G represent? _____

16. What kind of line is L? _____

CHAPTER 9

Practice Exercise 2

1. Put an A on the arrow side of the reference line for the following welding symbols.

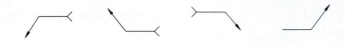

2. Put an O on the other side of the reference line for the following welding symbols.

3. Circle the incorrectly drawn welding symbols.

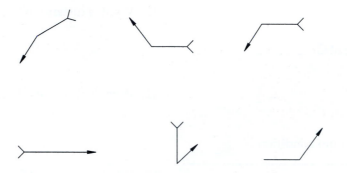

For questions 4 through 10, describe completely the requirements given by the following welding symbols.

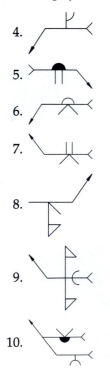

4.

5.

6.

7.

8.

9.

10.

CHAPTER 9
Math Supplement

U.S./Metric Length Conversions

Note: This chapter math supplement is limited to some of the more common conversions that may be required when you are working with some welding fabrication drawings. Two numbers, when memorized and used properly, provide the most convenience when converting between metric and U.S. standard length conversions. These two numbers are:

25.4

and:

.03936

The number 25.4 is important because that is how many millimeters are in 1 inch. The number .03936 is important because that is how many inches are in 1 millimeter. The conversion methods are as follows:

<u>To convert from:</u>

Inches to millimeters (mm): Multiply the number of inches by 25.4.

Millimeters (mm) to inches: Multiply the number of millimeters by .03936.

Inches to centimeters (cm): Multiply the number of inches by 2.54.

Centimeters (cm) to inches: Multiply the number of centimeters by .3936.

Inches to meters (m): Multiply the number of inches by .0254.

Meters (m) to inches: Multiply the number of meters by 39.36.

Math Supplement Practice Exercise 1

Convert the following. Write your answer in the space provided.

1.	50 mm to inches	
2.	14″ to millimeters	
3.	30 cm to inches	
4.	250″ to meters	
5.	3.6 meters to inches	
6.	63″ to centimeters	
7.	45.5 cm to inches	
8.	175 mm to inches	
9.	$\frac{3''}{4}$ to mm	
10.	1′8″ to cm	
11.	0.625″ to cm	
12.	22.5 mm to inches	
13.	$16\frac{3''}{4}$ to mm	
14.	12.8 meters to feet and inches	
15.	256 cm to feet and inches	

Advanced Welding Symbols

OBJECTIVE

- Be able to interpret welding symbols that include all of the information that could be used on them

KEY TERMS

Weld all around	Pitch	Spacer
Field weld	Chain intermittent weld	Convex contour
Weld length	Staggered intermittent weld	Concave contour
Intermittent weld	Consumable insert	Flush contour
Skip weld	Backing	Melt-through

OVERVIEW

Many additional elements can be added to the basic parts of the welding symbol. The additional elements and their placement, along with the basic parts of the welding symbol, are shown in Figures 10-1 and 10-2. They are described throughout the remainder of this chapter.

WELD ALL AROUND

The specification to **weld all around** requires that the weld be made to encapsulate the entire joint. In the case of a circular joint, the weld all around symbol is not required. The weld all around symbol consists of a circle that is placed over the intersection where the end of the reference line meets the arrow. Examples of weld all around welds and welding symbols are shown in Figures 10-3 and 10-4.

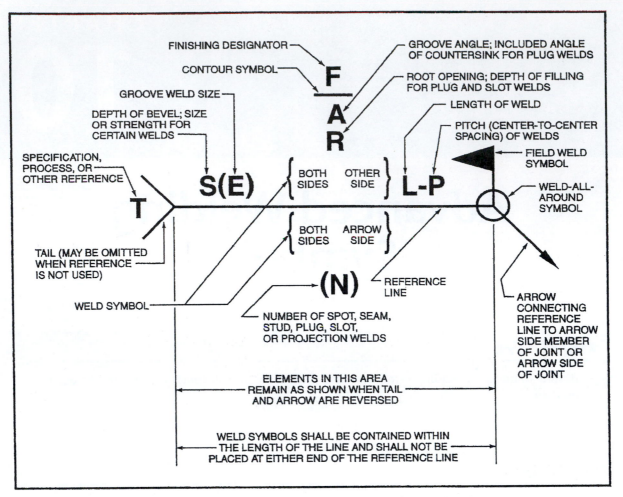

Figure 10-1 Standard Locations of the Elements of a Welding Symbol (*AWS A2.4:2007, Figure 3 reproduced and adapted with permission from the American Welding Society (AWS), Miami, FL.*)

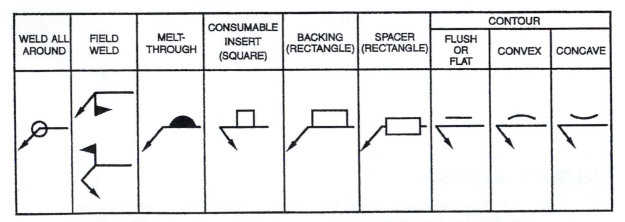

Figure 10-2 Supplementary Symbols (*AWS A2.4:2007, Figure 2 reproduced and adapted with permission from the American Welding Society (AWS), Miami, FL.*)

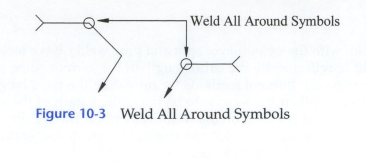

Figure 10-3 Weld All Around Symbols

PLATE WELDED ON TOP OF ANOTHER PLATE

CHANNEL WELDED TO A PLATE

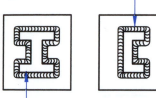

BEAM WELDED TO A PLATE

Figure 10-4 Example of Weld All Around Welds

FIELD WELD

A **field weld** is defined by the American Welding Society (AWS) as "[a] weld made at a location other than a shop or the place of initial construction."[1] The field weld symbol consists of a flag that is placed at the intersection where the end of the reference line meets the arrow (see Figure 10-5).

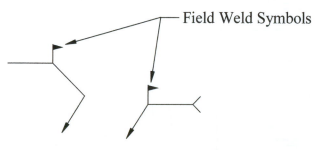

Field Weld Symbols

Figure 10-5 Field Weld Symbol Examples

[1] American Welding Society (AWS) A2 Committee on Definitions and Symbols, AWS A3.0:2001 Standard Welding Terms and Definitions, 14.

WELD LENGTH

Each weld, with the exception of spot and plug welds, has a length component. The **weld length** may be the entire length of the joint or some portion thereof. There are several different methods for providing the weld length information on a drawing. When the weld is to be the entire length of the joint, the length component is not required on the welding symbol. The welding symbol points to the joint requiring the weld, and the weld is made the entire length of that particular joint, as shown in Figure 10-6.

When the weld length is not required to extend the complete length of the joint, it can be defined by placing the required length to the right of the weld symbol. The welding symbol then points to the area of the joint requiring the weld. It may replace the standard length dimension, as shown in Figure 10-7.

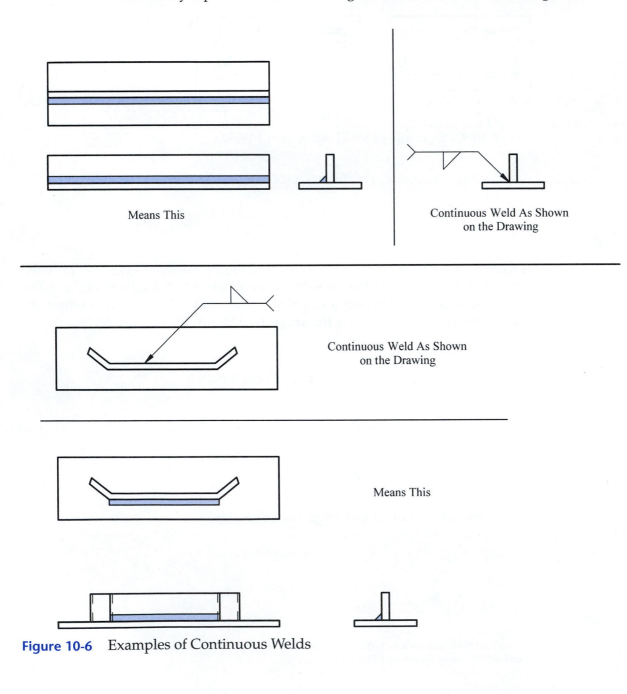

Means This

Continuous Weld As Shown
on the Drawing

Continuous Weld As Shown
on the Drawing

Means This

Figure 10-6   Examples of Continuous Welds

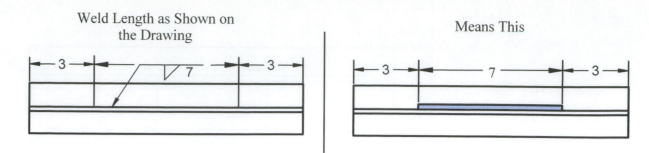

Weld Length as Shown on the Drawing

Means This

Figure 10-7   Weld Length Specified on Welding Symbol Between Extension Lines

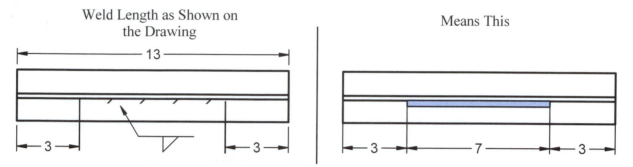

Weld Length as Shown on the Drawing

Means This

Figure 10-8   Weld Length Specified on Welding Symbol Between Extension Lines with Section Lines Representing the Weld Area

Placing section lines in the area where the weld is to be placed can also be used in combination with standard dimensions and welding symbols to identify the required weld length and weld location. See Figure 10-8.

INTERMITTENT WELDS

An **intermittent weld**, also called a **skip weld**, consists of a series of welds placed on a joint, with unwelded spaces between each of the welds. The individual weld segments in an intermittent weld have a length and **pitch** component. The weld length is the linear distance of each weld segment. The length is shown in the welding symbol to the right of the weld symbol. The pitch is the center-to-center distance of each of the weld segments. It is shown to the right of the length on the welding symbol, with a dash between the two. This concept is shown in Figure 10-9.

When intermittent welds are placed on both sides of a joint, they can be either directly opposite each other, known as a **chain intermittent weld** or they can be offset, known as a **staggered intermittent weld**. The chain intermittent weld is shown in Figure 10-10. The staggered intermittent weld is shown in Figure 10-11.

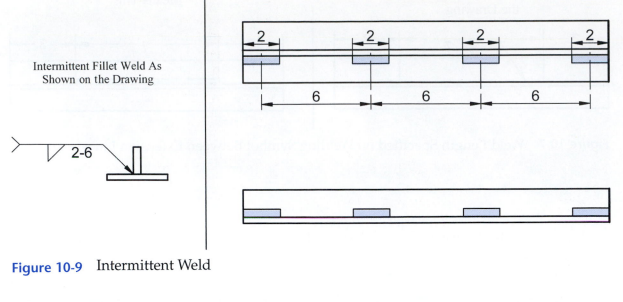

Intermittent Fillet Weld As
Shown on the Drawing

Means This

Figure 10-9 Intermittent Weld

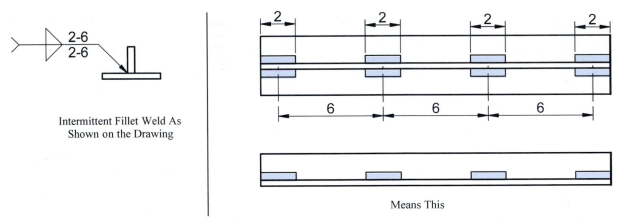

Intermittent Fillet Weld As
Shown on the Drawing

Means This

Figure 10-10 Chain Intermittent Weld

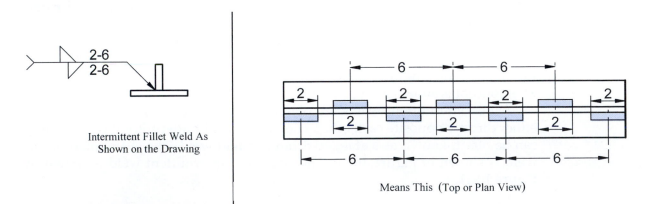

Intermittent Fillet Weld As
Shown on the Drawing

Means This (Top or Plan View)

Figure 10-11 Staggered Intermittent Weld

WELD CONTOUR SYMBOLS

The contour of a weld refers to the shape of its face. A weld with a **convex contour** has a face that protrudes out in a convex shape from its toes; a weld with a **concave contour** has a face that is concave, or sinks in from its toes; and a weld with a **flush contour** has a face that is flush with the base metal, or is flat from one toe to the other. When required, the contour symbol is added to a welding symbol so that it is oriented to mimic the required contour of the weld (see Figure 10-12). When a contour symbol is not added to the welding symbol, standard welding and shop practices should be followed.

FINISH SYMBOLS

Placing a finish symbol adjacent to the contour symbol specifies the method of making the contour. The finish symbols are made up of letters. The letters and their corresponding methods are listed below.

U = **unspecified** ➤ This means that any appropriate method may be used.
G = **grinding**
M = **machining**
C = **chipping**

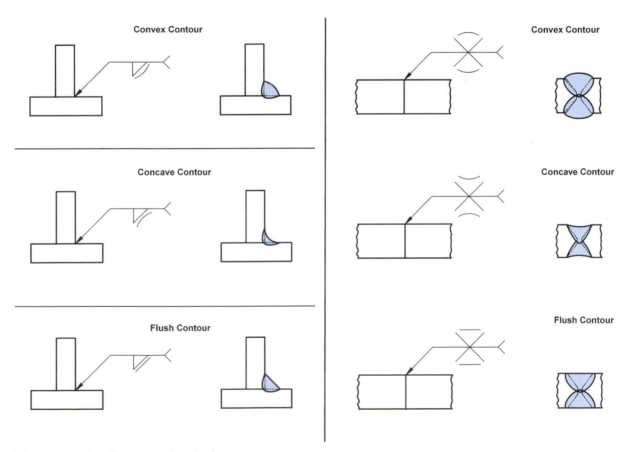

Figure 10-12   Contour Symbols

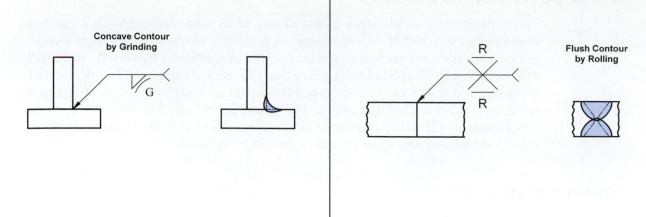

Figure 10-13   Examples of Contour Symbols with Method of Finish

R = **rolling**
H = **hammering**
P = **planishing**

See Figure 10-13 for an example of a finish Symbol.

FILLET WELDS

A welding symbol for a fillet weld includes the required fillet weld symbol and (as needed) the size, length, pitch, contour, method of making the contour, weld all around, field weld, and any other supplemental information listed in the tail of the welding symbol. See Figure 10-14.

The size of the fillet weld is shown to the left of the weld symbol. It represents the length of the legs of the largest right triangle that can fit within the joint, with the vertex of the triangle located at the intersection of the two members being joined. See Figure 10-15.

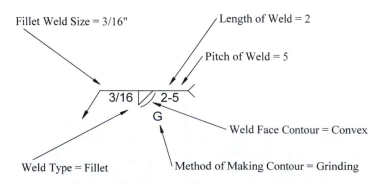

Figure 10-14   Example of a Welding Symbol for a Fillet Weld

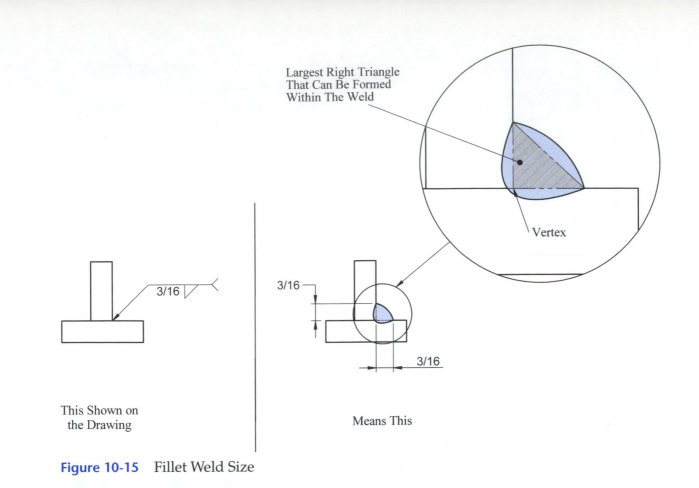

Figure 10-15   Fillet Weld Size

UNEQUAL LEG FILLET WELDS

A fillet weld can be required to have unequal legs. In such cases, the size for each of the legs is shown on the welding symbol to the left of the weld symbol and is written in parentheses. The only way to know which leg goes with which size is through either a detail drawing that shows the weld joint, as in Figure 10-16; a note; or other revealing information, such as one of the legs is required to be longer than one of the sides, as in Figure 10-17.

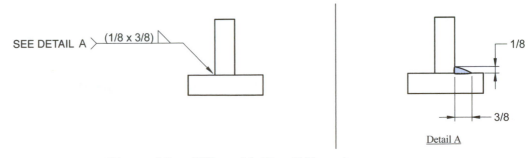

Figure 10-16   Unequal Leg Fillet with Detail Drawing

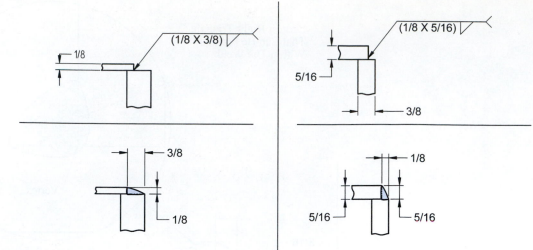

Figure 10-17 Unequal Leg Fillets that could be Shown Without Detail

GROOVE WELDS

The components of a welding symbol for a groove weld include the required groove weld symbol for the type of groove weld called for and may also include, as needed, the size, depth of preparation, root opening, groove angle (also called included angle) contour, method of making the contour, length, pitch, weld all around, field weld, and any other supplemental information listed in the tail of the welding symbol. See Figures 10-18 and 10-19.

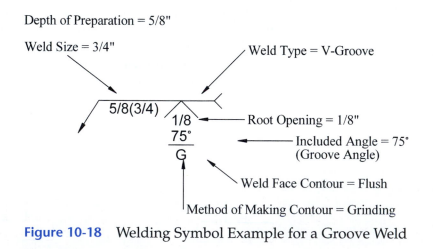

Figure 10-18 Welding Symbol Example for a Groove Weld

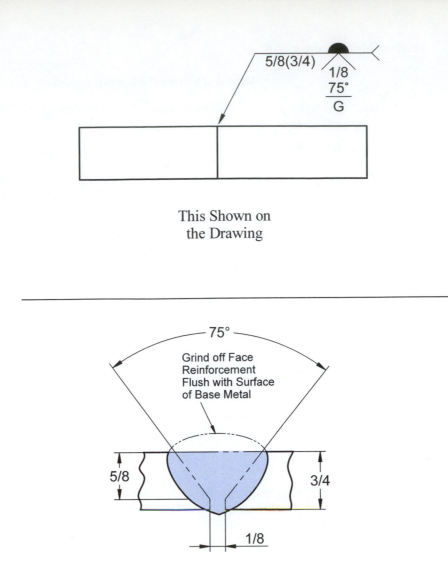

This Shown on
the Drawing

Means This

Figure 10-19   Example of a Groove Weld

MELT-THROUGH

Melt-through is defined by the American Welding Society (AWS) as "visible root reinforcement produced in a joint welded from one side."[2] In other words, melt-through is the penetrated weld metal that extends beyond the base metal on the back side of a joint welded from the opposite side. Height, contour, method of contour, and tail specifications are all supplemental types of information that can be added to the melt-through weld symbol. See Figures 10-20 and 10-21.

---

[2] AWS A3.0:2001 Standard Welding Terms and Definitions, 25.

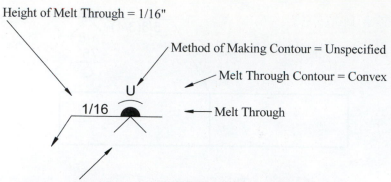

Height of Melt Through = 1/16"

Method of Making Contour = Unspecified

Melt Through Contour = Convex

1/16   U

Melt Through

Melt Through Symbols Will Have Some Type of Weld
Symbol on the Other Side of the Reference Line

Figure 10-20   Welding Symbol with Melt Through

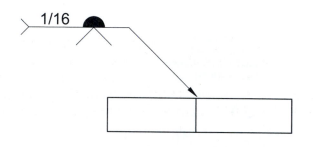

1/16

This Shown on the
Drawing

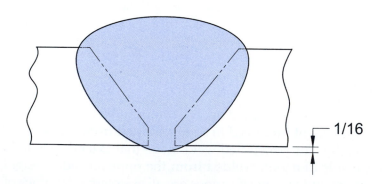

1/16

Means This

Figure 10-21   Melt Through Example

CONSUMABLE INSERT

A **consumable insert** is preplaced filler metal that is fused into the root of the joint. Consumable inserts come in five different classes. The five classes correlate to the following 5 shapes:

1. Class 1: Inverted T cross section
2. Class 2: J shaped cross section
3. Class 3: Solid ring shape
4. Class 4: Y shaped cross section
5. Class 5: Rectangular shaped cross section

The class number (1, 2, 3, or 5) required for the application should be shown in the tail of the welding symbol, as in Figure 10-22. AWS A5.30/A5.30M:2007 Specification for Consumable Inserts further defines and specifies the requirements of the five classes of consumable inserts.

BACKING

Backing refers to placing something against a weld joint to support the molten metal. In most cases, it is placed against the back side of the weld joint, thus the term *backing*. For certain instances, however, like electroslag and electrogas welding, it can be used on both sides of the weld joint because both sides of the joint have molten metal at the same time and both sides must be supported. There are many types of backing, including backing bar or backup bar, backing pass, backing weld, backing ring, backing shoe, and backing strip, and many methods for applying backing. Backing can be made from material that will fuse

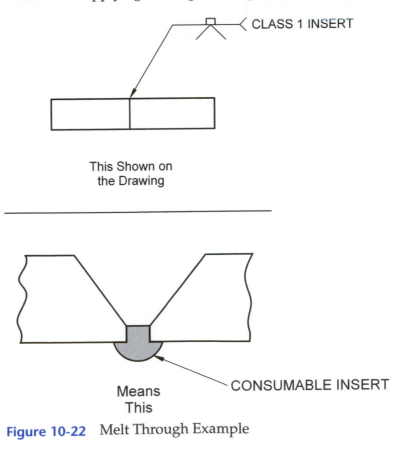

Figure 10-22   Melt Through Example

into the weld or it can be made from material that will not fuse into the weld. Fused backing can be required to be removed after welding or left on to become part of the completed weld joint. When backing is to be removed after welding, an "R" is placed within the perimeter of the backing symbol. Figure 10-23 shows a backing symbol with an "R", indicating backing material that is to be removed after welding. Figure 10-24 shows a welding symbol and weld for a joint that is

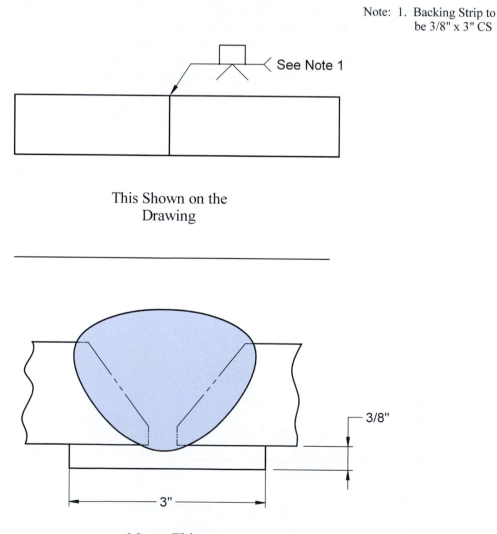

Backing Material
R = Remove After Welding

Backing Symbols Will Have a Weld Symbol on the
Opposite Side of the Reference Line as the Backing Symbol.

Figure 10-23   Welding Symbol with Backing Material

Note: 1. Backing Strip to
be 3/8" x 3" CS

See Note 1

This Shown on the
Drawing

3/8"

3"

Means This

Figure 10-24   Backing Strip Weld

to have a backing strip that is to be left on after welding. Notice there is no "R" used in this symbol.

SPACER

For certain applications, a spacer can be used between weld joint members. The requirement to use a spacer is shown by the spacer symbol, which is a rectangular box placed on the center of the reference line, as shown in Figure 10-25. When the spacer symbol is used, the size and type of material should be shown in the tail of the welding symbol.

BACKING WELDS AND BACK WELDS

Backing welds and back welds use the same weld symbol, which looks like an unshaded half circle. Determination of which type (backing weld versus back weld) is required based on the symbol alone cannot be made. A note in the tail of the welding symbol may be provided to specify which type of weld is required, a multiple reference line may be used to indicate the sequence of operation, or it may be specified in the welding procedure. Height, contour, and method of contour are all types of information that can be added to the back or backing weld symbol. See Figure 10-26.

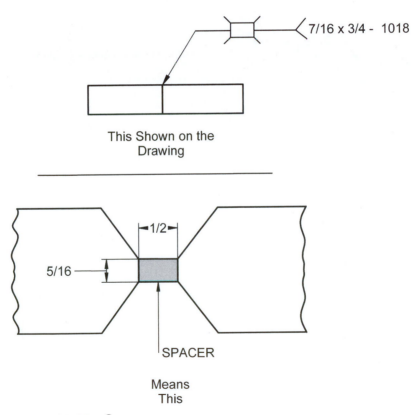

Figure 10-25 Spacer

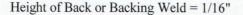
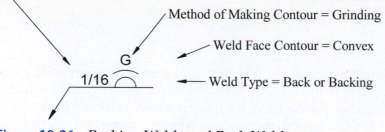

Height of Back or Backing Weld = 1/16"

Method of Making Contour = Grinding

Weld Face Contour = Convex

Weld Type = Back or Backing

Figure 10-26   Backing Welds and Back Welds

EDGE WELDS

The welding symbol for an edge weld includes the edge weld symbol and when required, the following additional information: size, length, pitch, contour and method of finish. When the size of an edge weld is specified, it refers to the throat of the weld (the distance from the root to the face of the weld). Figure 10-27 shows the symbol for an edge weld with all of the supplemental information that may be used. Figure 10-28 shows the meaning of the weld size for an edge weld.

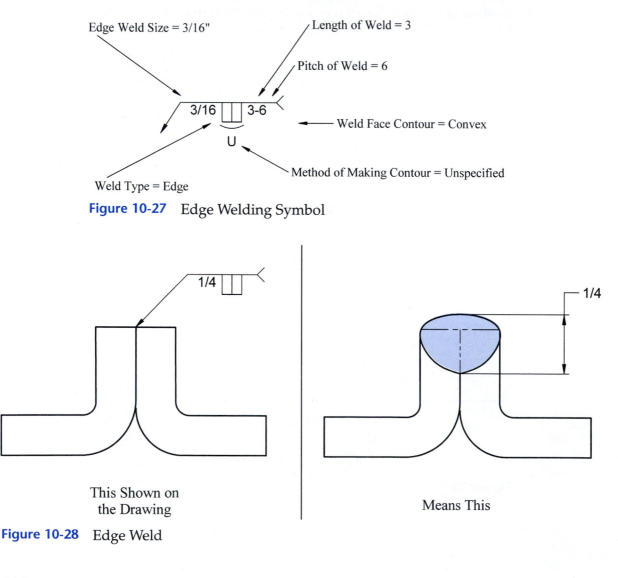

Edge Weld Size = 3/16"

Length of Weld = 3

Pitch of Weld = 6

Weld Face Contour = Convex

Method of Making Contour = Unspecified

Weld Type = Edge

Figure 10-27   Edge Welding Symbol

This Shown on
the Drawing

Means This

Figure 10-28   Edge Weld

SURFACING WELDS

The size of a surfacing weld is determined by its height from the substrate to the face of the weld. The weld direction of surfacing welds is identified in the tail of the welding symbol by the terms *axial, circumferential, longitudinal,* and *lateral,* or it may be identified in a welding procedure. See Figure 10-29. Examples of surfacing weld placement direction are shown in Figure 10-30.

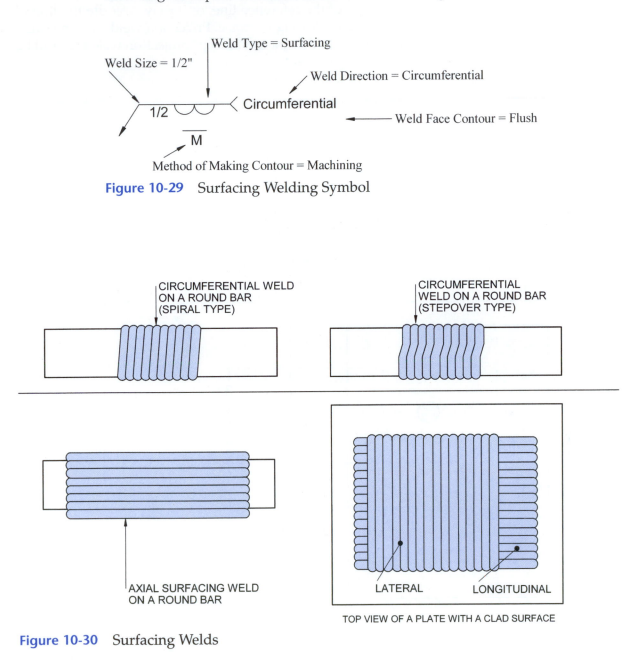

Weld Type = Surfacing

Weld Size = 1/2"

Weld Direction = Circumferential

1/2 Circumferential

Weld Face Contour = Flush

M

Method of Making Contour = Machining

Figure 10-29 Surfacing Welding Symbol

CIRCUMFERENTIAL WELD ON A ROUND BAR (SPIRAL TYPE)

CIRCUMFERENTIAL WELD ON A ROUND BAR (STEPOVER TYPE)

AXIAL SURFACING WELD ON A ROUND BAR

LATERAL   LONGITUDINAL

TOP VIEW OF A PLATE WITH A CLAD SURFACE

Figure 10-30 Surfacing Welds

SPOT AND PROJECTION WELDS

Figure 10-31 shows how the size, quantity, **pitch**, and process for spot welds are depicted. The size refers to the size of the weld at the junction of the faying surfaces of the materials being joined. Shear strength, given in pounds or newtons, may be used in place of the size dimension. There is no length dimension. As you saw in Chapter 9, the spot weld symbol may be placed on the arrow side or other side of the reference line, or it may straddle the line to indicate no arrow side or other side significance. Projection welds use the same symbol, size, quantity, and pitch designators, except projection welds should be shown only as arrow side or other side.

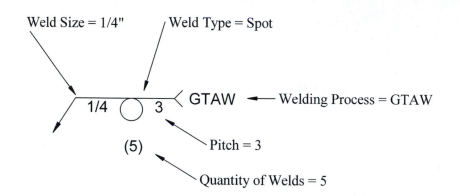

Weld Size = 1/4"   Weld Type = Spot

1/4 ◯ 3 ⟨ GTAW ◀— Welding Process = GTAW

(5)   Pitch = 3

Quantity of Welds = 5

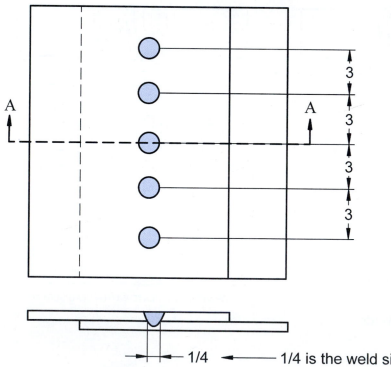

1/4 ◀———— 1/4 is the weld size at the faying surfaces

Section A-A

Figure 10-31 Welding Symbol Example for a Spot or Projection Weld

SEAM WELDS

The seam-welding symbol indicates a weld that takes place between the faying surfaces of a lap joint that may be composed of two or more lapped pieces. The weld is done typically by moving the lap joint between two rolling electrical contact wheels that pass current through the joint to create a type of rolling spot weld called a resistance seam weld (RSEW), or by a welding process that has enough arc energy to melt through one of the plates and weld down into the other(s). Seam welds can be made by making overlapping spot welds to form a seam. Figure 10-32 shows two different seam welding symbols with supplemental information that may be used.

Figure 10-33 shows two different ways that seam welds can be made: end to end or side by side. End to end is considered the way indicated by the welding symbol unless a separate detail indicates otherwise.

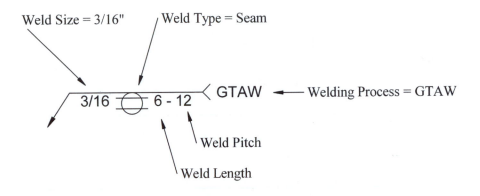

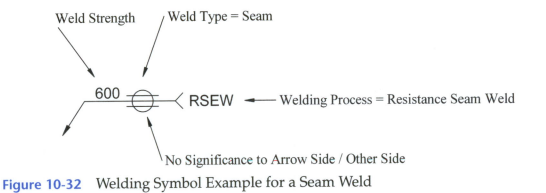

Figure 10-32 Welding Symbol Example for a Seam Weld

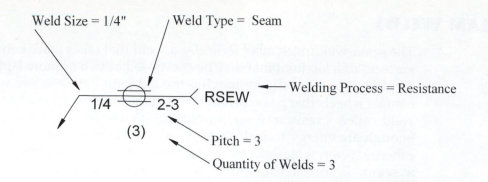

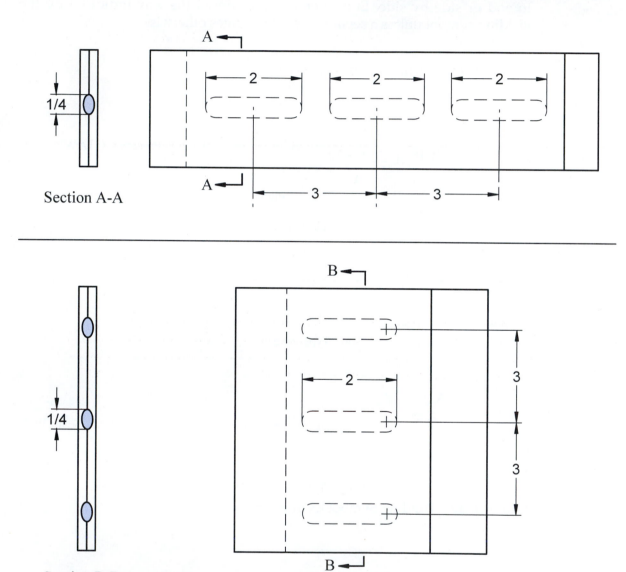

Section A-A

Section B-B

Figure 10-33 Different Configurations for Seam Welds

STUD WELDS

The basic information for stud welds is given in the welding symbol shown in Figure 10-34. Additional information on the specifics of the stud is provided, as necessary, by other means, such as a note, bill of material specification, or specification in the tail of the welding symbol.

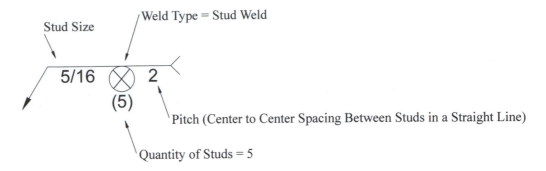

Note: Stud welds are only shown arrow side.

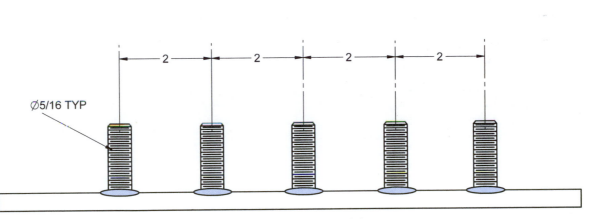

Figure 10-34   Welding Symbol Example for a Stud Weld

PLUG AND SLOT WELDS

Plug and slot symbols contain similar information except that plug welds do not have a length component and the size of a plug weld is its diameter. They may be completely filled or filled to a depth specified within the weld symbol. See Figure 10-35.

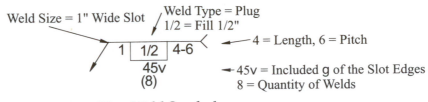

Figure 10-35   Plug Weld Symbol

CHAPTER 10
Practice Exercise 1

Describe completely all of the information contained in the following welding symbols.

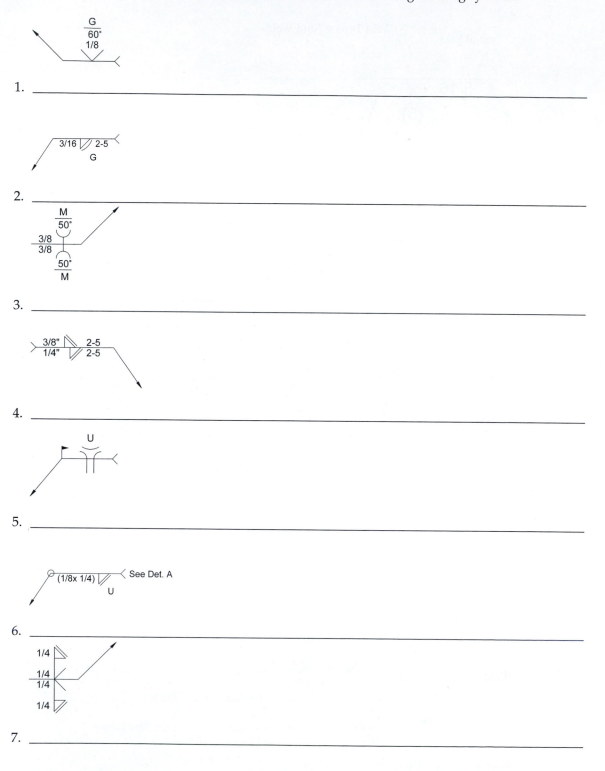

1. _____

2. _____

3. _____

4. _____

5. _____

6. _____

7. _____

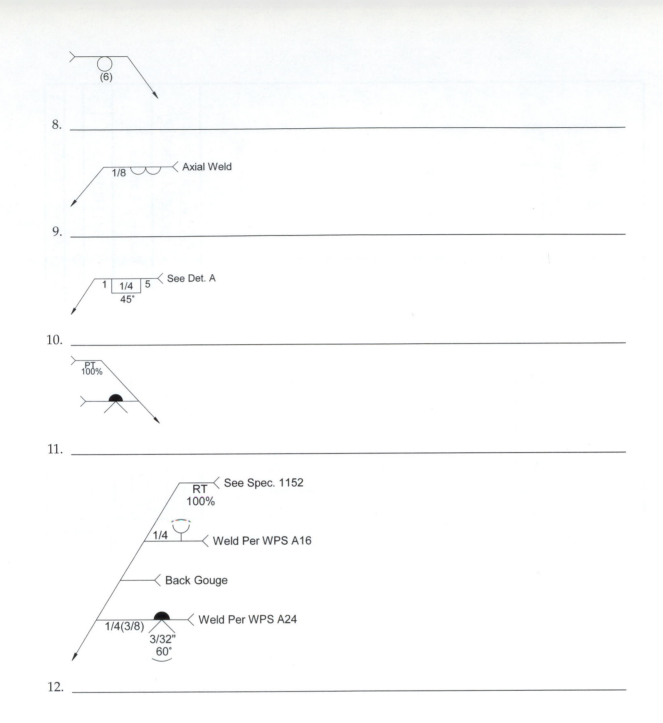

8. _____

9. _____

10. _____

11. _____

12. _____

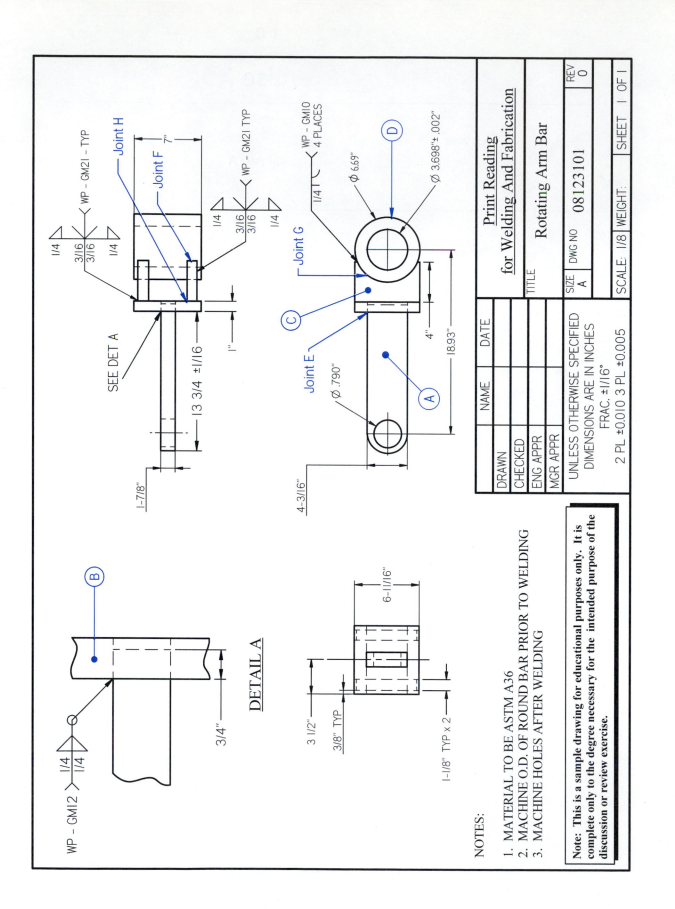

DETAIL A

NOTES:

1. MATERIAL TO BE ASTM A36
2. MACHINE O.D. OF ROUND BAR PRIOR TO WELDING
3. MACHINE HOLES AFTER WELDING

Note: This is a sample drawing for educational purposes only. It is complete only to the degree necessary for the intended purpose of the discussion or review exercise.

Print Reading
for Welding And Fabrication

	NAME	DATE
DRAWN		
CHECKED		
ENG APPR		
MGR APPR		

UNLESS OTHERWISE SPECIFIED DIMENSIONS ARE IN INCHES FRAC. ±1/16° 2 PL ±0.010 3 PL ±0.005

TITLE   Rotating Arm Bar

SIZE A	DWG NO	08123101	REV 0

SCALE: 1/8 | WEIGHT: | SHEET 1 OF 1

Practice Exercise 2

Refer to Drawing Number 08123101 on page 172.

1. What is the minimum and maximum acceptable length of part A? Minimum _____
   Maximum _____

2. Who approved this drawing? _____

3. Describe completely the weld required at joint F. _____

   _____

   _____

4. What is the MMC and LMC of the I.D. of the round bar?
   MMC _____   LMC _____

5. When are the holes to be machined? _____

6. What is the dimensioned thickness of part C? _____

7. Draw the welding symbol for the weld required at joint H.

8. Describe completely the weld required at joint E. _____

   _____

   _____

9. What is the maximum overall acceptable size of the rotating arm bar? _____

10. Describe completely the weld required at joint G. _____

    _____

    _____

11. What is the center-to-center distance between the two holes? _____

11. What is the specification for the required material? _____

12. What welding procedure is to be used at joint H? _____

13. How long is the round bar? _____

14. What are the dimensions of part B? _____

15. What is the overall maximum acceptable length of the rotating arm bar? _____

CHAPTER 10
Math Supplement

Working with Drawing Scales

Drawings are either made to conform to a particular ratio between the length of lines on the drawing to the actual part sizes or they are not. If they conform to a particular ratio, they are drawn to scale. If they do not conform to a particular ratio, they are said to be not to scale.

Drawings that are drawn to scale are made either smaller than the actual object, which is called **reduced scale;** the same size as the actual object, called **full scale;** or larger than the actual object, called **enlarged scale.** The scale used is typically given within the title block area of the drawing. This scale covers the entire drawing except for any localized scales that are noted for certain individual views or other specific areas of the drawing.

Drawing scales (ratios) are normally written with an equals sign, such as 3/8" = 1" but they may also be written as a ratio like 1:1, or simply written with words such as half scale or whole scale. Half scale is the same as ½ = 1 or simply ½ scale, and whole scale is the same as 1 = 1 or 1:1.

- Typical reduced scale sizes are: 1/8" = 1", 1/4" = 1", 3/8" = 1", 1/2" = 1", 3/4" = 1".
- Typical enlarged scale sizes 2:1, 3:1, 4" = 1", etc.

The number to the left of the equals sign is the length drawn to represent the distance of the true-life length of the object for every unit of the amount shown to the right of the equals sign. For example, a scale of 1/8" = 1" means that 1/8" is shown on the drawing for every 1" of the object's true size. The object's true size is designated on the drawing with the dimension that is shown.

Finding the scale factor for a drawing

Estimating a non-dimensioned measurement on a drawing or mathematically determining the required length of a line required on a drawing for many scales such as 1/2" = 1" or 2" = 1" is easy to figure out just by looking at the scale or doing simple math. Other problems can be more difficult. An easy way to work with any scale is to first determine the scale factor. This can be accomplished by using the following steps.

Step 1: Make sure the units (inches, mm, feet, etc.) on both sides of the equals sign are the same.
Step 2: Invert the number (move the numerator to the denominator and the denominator to the numerator) that is on the left side of the equals sign. *Note:* For the number 1, this step can be omitted.
Step 3: Multiply the inverted number by the number to the right of the equals sign. The answer to this step is the scale factor.

Example 1: Determine the scale factor for a drawing scale of $\frac{3}{8}$" = 1'.

Step 1: Change 1' to 12' so that the units on both sides of the equals sign are the same.
Step 2: Invert the number that is on the left side of the equals sign so that $\frac{3}{8}$ becomes $\frac{8}{3}$.
Step 3: Multiply $\frac{8}{3}$ times 12 for an answer of $\frac{96}{3}$ which reduces to the final answer of 32.

Determining the length of a dimensioned line on a drawing by using a scale factor

Step 1: Determine the scale factor by following the steps outlined in example 1.
Step 2: Divide the given dimension by the scale factor. The result is the length of the line on the drawing that represents the dimensioned distance.

Example 2: Find the length of a line dimensioned 36" on a drawing that has a scale of $\frac{1}{16}'' = 1''$.

Step 1: Determine the scale factor of 16 by using the steps outlined in example 1.
Step 2: Divide the dimensioned distance of 36 by the scale factor of 16 to achieve the final answer of $2\frac{1}{4}''$.

Estimating a non-dimensioned distance on a drawing

When the estimate of a non-dimensioned distance on a drawing is needed it can be found by using the following steps.

Step 1: Determine the scale factor by following the steps outlined in example 1.
Step 2: Measure the distance on the drawing.
Step 3: Multiply the measured distance by the scale factor. The result is the estimated distance.

Note: Errors due to print copying reductions or enlargements and ruler placement during measurement can occur when making estimations based on scales. The dimensions given on the drawing should be used to calculate non-dimensioned areas of the drawing when performing fabrication or other activities that require non-estimated measurements.

Example 3: Estimate a non-dimensioned distance that measures 6" on a print that is drawn $\frac{1}{4}''$ scale.

Step 1: Determine the scale factor of 4 by using the steps outlined in example 1.
Step 2: Multiply the scale factor of 4 by the 6-inch measurement to achieve the final answer of 24 inches.

CHAPTER 10

Math Supplement Practice Exercise 1

Find the length of the line that is shown on the drawing to represent the given distance.

1. A line dimensioned at 6″ is shown on a drawing with a scale of $\frac{3}{8}$ = 1. What is the length of the line?

2. A line dimensioned at 17″ is shown on a drawing with a scale of $\frac{1}{16}$ = 1. What is the length of the line?

3. A line dimensioned at $4\frac{3}{16}$″ is shown on a drawing with a $\frac{1}{4}$ scale. What is the length of the line?

4. A line dimensioned at $17\frac{1}{2}$″ is shown on a drawing with a scale of $\frac{5}{8}$ = 1. What is the length of the line?

5. A line dimensioned at $1\frac{1}{2}$″ is shown on a drawing with a scale of 4:1. What is the length of the line?

6. Estimate a non-dimensioned distance that measures 3″ on a drawing that has a scale of $\frac{1}{2}$″ = 1′.

7. Estimate a non-dimensioned distance that measures 7″ on a drawing that has a scale of $\frac{1}{8}$″ = 1″.

8. Estimate a non-dimensioned distance that measures $9\frac{1}{2}$″ on a drawing that has a scale of 1″ = 1′.

9. Estimate a non-dimensioned distance that measures $6\frac{1}{4}$″ on a drawing that has a scale of 2 : 1.

10. Estimate a non-dimensioned distance that measures $5\frac{3}{16}$″ on a drawing that has a scale of $\frac{1}{4}$″ = 1″.

Additional Views

- Be able to identify the different kinds of views shown in this chapter
- Be able to interpret and understand the views shown in this chapter

KEY TERMS

Full section	Revolved section	Broken-out section
Half section	Removed section	True orientation
Offset section	Auxiliary view	
Aligned view	Auxiliary section	

OVERVIEW

A design may include a number of views in addition to the basic views discussed in Chapter 3. These additional views further clarify the drawing requirements, and they are described below.

FULL SECTION

A **full section** shows the object from the viewpoint of a theoretical cut through the object from one side to the other. Cutting plane lines are used to indicate where the object in the view is to be theoretically cut. The arrows of the cutting plane line point in the viewing direction of the section view. Identification of the section view is done via capital letters placed at the endpoints of the arrows from the cutting plane lines. An example of a full section is shown in Figure 11-1.

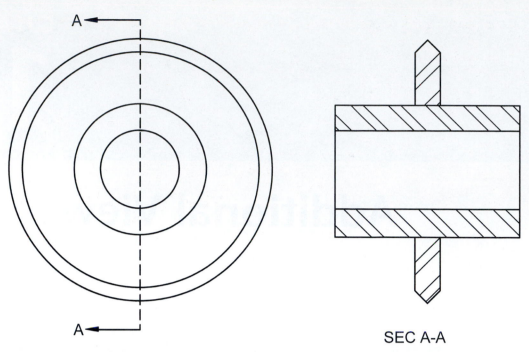

SEC A-A

Figure 11-1   Full Section

HALF SECTION

A **half section** shows an object from the viewpoint of a theoretical cut halfway through the object, essentially removing a quarter of the object. A half section is shown in Figure 11-2.

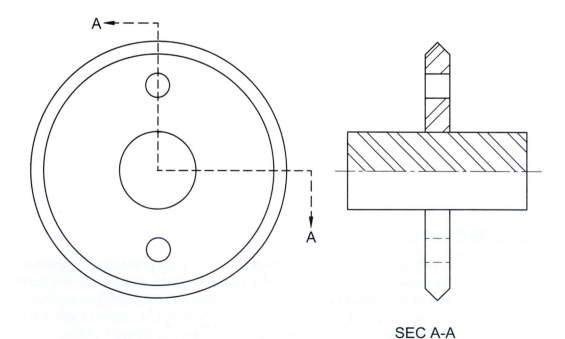

SEC A-A

Figure 11-2   Half Section

OFFSET SECTION

An **offset section** is where the cutting plane line does not go straight through the object, but instead is offset to show the features that the engineer wants in the section view. An offset section is shown in Figure 11-3.

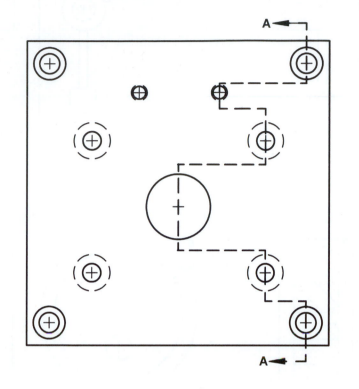

SECTION A-A

Figure 11-3   Offset Section

ALIGNED VIEW

An **aligned view** is used to show the view of an angled feature of a part. The viewing or cutting plane line shows how the part should be viewed. The aligned view shows the part as if it were rotated into alignment. Figures 11-4 and 11-5 each show a standard view of an object in View A-A and an aligned view of the same object in View B-B.

REVOLVED SECTION

When a section is placed at right angles within the object or between short break lines within the object, it is called a **revolved section**. It shows the shape and material type of the object without taking up additional room on the drawing. A revolved section is shown in Figure 11-6.

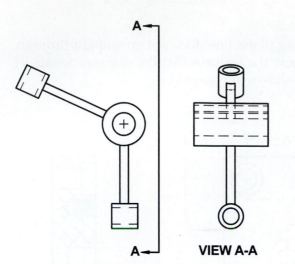

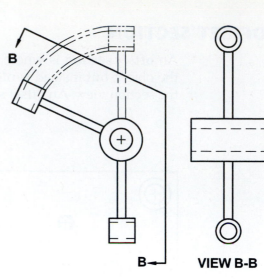

VIEW A-A

VIEW B-B

Figure 11-4   Standard View vs. Aligned View

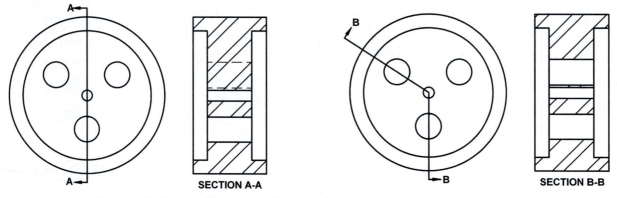

SECTION A-A

SECTION B-B

Figure 11-5   Standard Section View vs. Aligned Section View

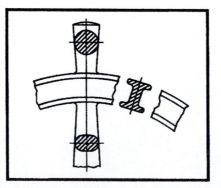

Figure 11-6   Revolved Sections *(Reprinted with modifications by the author from ASME Y14.3-2003 by permission of The American Society of Mechanical Engineers. All rights reserved.)*

REMOVED SECTION

A **removed section** is a section that has been removed and placed outside the object, as shown in Figure 11-7.

Figure 11-7 Removed Section *(Reprinted with modifications by the author from ASME Y14.3-2003 by permission of The American Society of Mechanical Engineers. All rights reserved.)*

MULTIPLE SECTION VIEWS

An object or an assembly can have more than one section view associated with it. See Figure 11-8 for an example of an object with more than one section view.

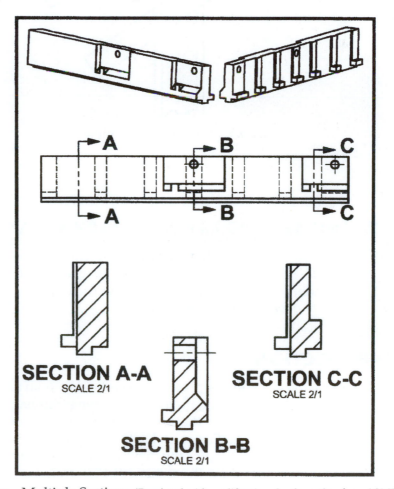

Figure 11-8 Multiple Sections *(Reprinted with modifications by the author from ASME Y14.3 2003 by permission of The American Society of Mechanical Engineers. All rights reserved.)*

AUXILIARY VIEW

An **auxiliary view** is placed perpendicular to an angled feature of another view in an orthographic projection drawing. It is used to show the true size and shape of a slanted plane so that the features of the object can be more clearly detailed and understood. See Figure 11-9 for an example.

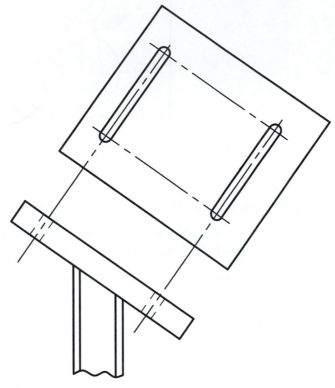

Figure 11-9   Auxiliary View

PRIMARY AND SECONDARY AUXILIARY VIEWS

When required, more than one auxiliary view can be shown in a drawing. In Figure 11-10, you can see two principal views, two primary auxiliary views, and a secondary auxiliary view. Notice how each of the auxiliary views is aligned with the view that it originates from.

AUXILIARY SECTION

An **auxiliary section** is placed perpendicular to an angled feature of another view in an orthographic projection drawing. Similar to an aligned view, it is placed in a drawing so that the additional details of internal features of the object can be shown. Figure 11-11 shows a partial view (indicated by the short break line at the bottom of the view) of an assembly, along with two auxiliary section views that show the sectioned object from two different perspectives.

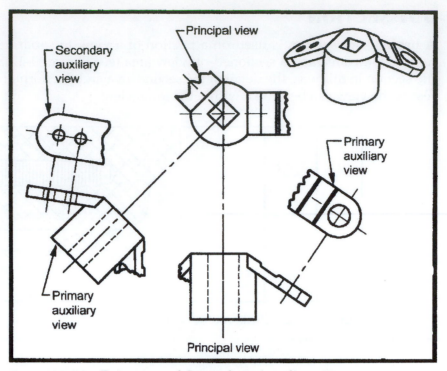

Figure 11-10 Primary and Secondary Auxiliary Views *(Reprinted with modifications by the author from ASME Y14.3-2003 by permission of The American Society of Mechanical Engineers. All rights reserved.)*

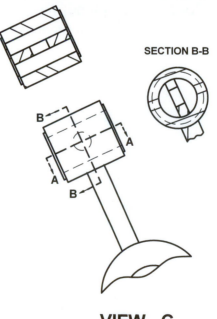

VIEW - C

Figure 11-11 Auxiliary Section

BROKEN-OUT SECTION

A **broken-out section** is used on a portion of a view to separate between a normal view area and a sectioned-off view area (see Figure 11-12). By creating the section in this way, the detail of the section area and the normal detail of the rest of the view can be shown within the same view.

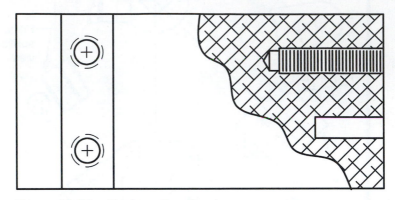

Figure 11-12   Broken Out Section

TRUE ORIENTATION

Figure 11-13 shows a tank with three shell sections, a head on each end, and six nozzles. The elevation view shows the nozzles that are located in the upper half of the tank rotated to the top of the view and the nozzles that are located in the bottom half of the tank rotated to the bottom of the view. This is done so that the nozzles can be shown at their full size and shape. The true orientation of the nozzles around the circumference of the tank is shown in the true orientation view.

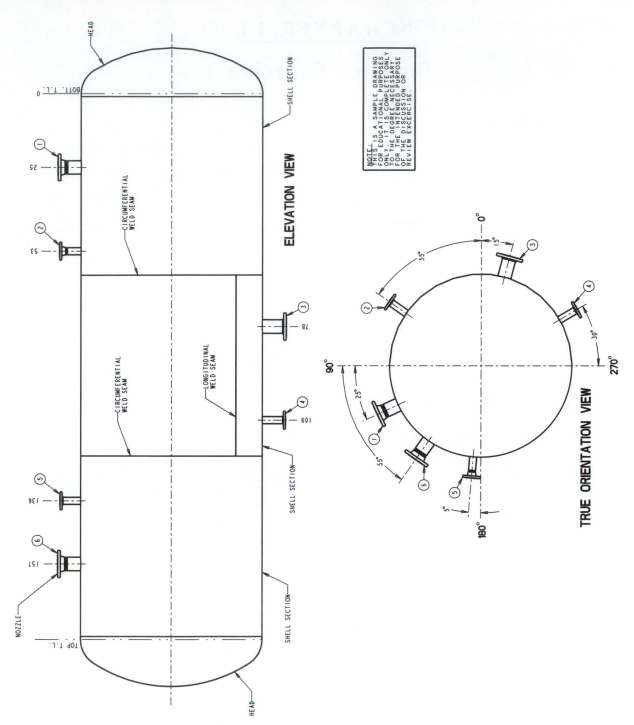

Figure 11-13 True Orientation View *(Courtesy of Kevin Schaaf.)*

CHAPTER 11

Practice Exercise 1

Identify the views shown below.

1. _____

2. _____

3. _____

4. _____

5. _____

6. _____

1.

2.

SECTION A-A

3.

SEC A-A

4.

SEC A-A

5.

6.

VIEW B-B

Math Supplement

Circles and Arcs

Circles

A circle can be defined as consisting of all points in a given plane that are equidistant from the center point (see Figure 11-14). The distance from the center point of a circle to the points that make up the circle is the radius. The distance from one side of a circle to the other side straight through the center is the diameter. A common formula for the radius of a circle is R = 1/2 D, where R is the radius and D is the diameter. A common formula for the diameter of a circle is two times the radius, written as D = 2R.

The circumference of a circle is the distance one time around its perimeter. See Figure 11-15. To find the circumference of a circle when you know the diameter, use the following formula:

$$C = \pi D$$

where C = circumference, the symbol π = 3.1416, and D = the diameter of the circle.

Example: The diameter of a circle is 10″. What is the circumference?
Answer: 31.4160″
Method: 3.1416 × 10″ = 31.4160″

To find the circumference of a circle when you know the radius, use the following formula:

$$C = 2\pi R$$

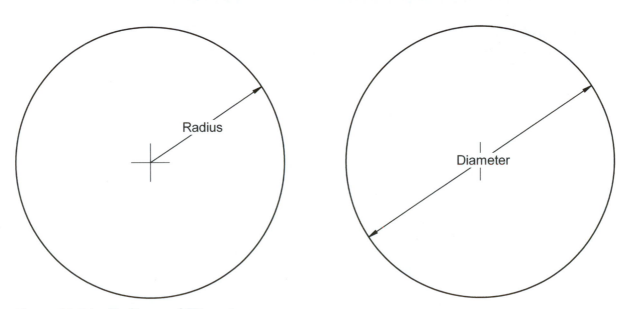

Radius

Diameter

Figure 11-14   Radius and Diameter

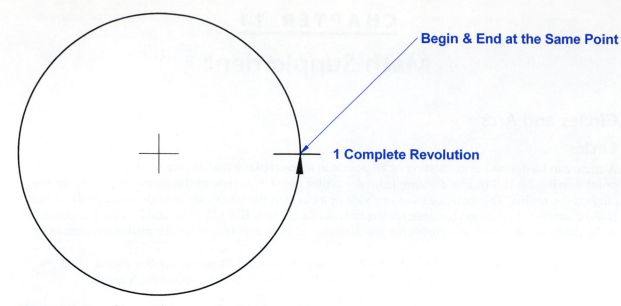

Begin & End at the Same Point

1 Complete Revolution

Figure 11-15   Circumference of a Circle

Example: The radius of a circle is 5″. What is the circumference?
Answer: 31.416″
Method: 2(3.1416)(5), which = 31.4160″

Notice that the answer is the same in both examples. This is because both use the same size circle, which has a radius of 5 and a diameter of 10. The two formulas are virtually the same, except that the D in the formula for the first example is replaced by 2R in the second example.

Circular Arcs

A circular arc is a portion of the circumference of a circle (see Figure 11-16).

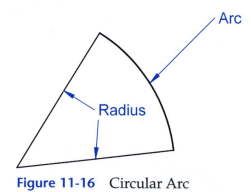

Arc

Radius

Figure 11-16   Circular Arc

CHAPTER 11

Math Supplement Practice Exercise 1

1. What is the diameter of a circle that has a radius of 10 inches? _____

2. What is the radius of a circle that has a diameter of 27.5 inches? _____

3. What is the circumference of the following circle? _____

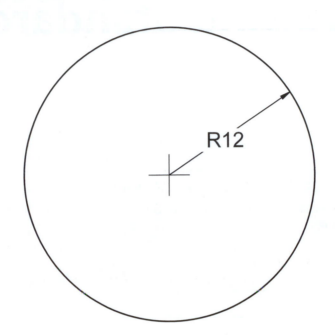

4. What is the circumference of the following circle? _____

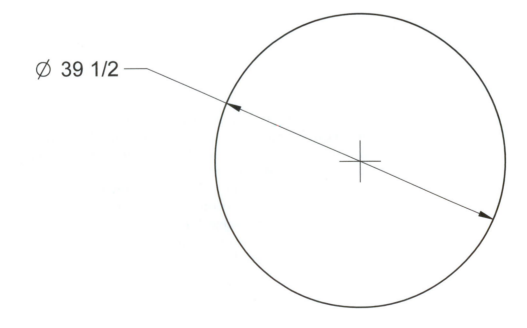

Drawing Standards

OBJECTIVES

- Understand what standards are and why they are important
- Understand the International Organization for Standardization (ISO) system for designating welding on drawings

KEY TERMS

Standard

American Welding Society (AWS)

American National Standards Institute (ANSI)

American Society of Mechanical Engineers (ASME)

International Organization for Standardization (ISO)

Broad root face V-groove

OVERVIEW

A **standard** can be defined as a standardized way of doing something, and it can be formal or informal. Formal standards have been documented and authorized by some authoritative body. Informal standards have not. Many standards cover how drawings are to be made. A small sample is listed in Table 12-1.

As you can see in Table 12-1, standards have been developed to define how lines, views, welding symbols, and so on, are to be made. More than one standard can be written on the same subject. For example, AWS A2.4:2007 and ISO 2553:1992 are both standards written by different organizations that cover how welds are to be represented on drawings. When reading a drawing, it is important to understand what standards, if any, were used while making the drawing so that it can be properly interpreted. Typically, most drawings from a

Table 12-1 Various Drawing Standards

Standard	From
ASME Y14.2–2008 Line Conventions and Lettering	American Society of Mechanical Engineers
ASME Y14.3–2003 Multiview and Sectional View Drawings	American Society of Mechanical Engineers
ASME Y14.5–2009 Dimensioning and Tolerancing	American Society of Mechanical Engineers
ASME Y14.24–1999 Types and Applications of Engineering Drawings	American Society of Mechanical Engineers
ASME Y14.34–2008 Associated Lists—Engineering Drawing and Related Documentation Practices	American Society of Mechanical Engineers
ASME Y14.36M–1996 Surface Texture Symbols	American Society of Mechanical Engineers
ASME Y14.100–2004 Engineering Drawing Practices	American Society of Mechanical Engineers
ISO 128-22:1999 Technical drawings—General principles of presentation—Part 22: Basic conventions and applications for leader lines and reference lines	International Organization for Standardization
ISO 128-34:2001 Technical drawings—General principles of presentation—Part 34 Views on mechanical engineering drawings	International Organization for Standardization
ISO 2553:1992 Welded, brazed, and soldered joints—Symbolic representation on drawings	International Organization for Standardization
AWS A2.4:2007 Standard Symbols for Welding, Brazing, and Nondestructive Examination	American Welding Society
AS 1100.201-1992 Technical drawing—Mechanical engineering drawing	Australian Standard
DIN 406-12 Engineering drawing practice; dimensioning; tolerancing of linear and angular dimensions	German Standard
BS 8888:2004 Technical Product Documentation (TPD) Specification for defining, specifying, and graphically representing products	British Standard

particular industry within an individual country follow similar overall concepts and guidelines that can be identified by an experienced reader. For example, most welding and fabrication drawings made in the United States typically follow **ASME/ANSI** standards for lines, views, lettering, and so on, and the

American Welding Society (AWS) Standard for Symbols for Welding, Brazing, and Nondestructive Examination. This does not mean that all drawings within an industry and country will be drawn the same way. Although standards exist, drawings from company to company and for each engineer can vary as to their quality and adherence to applicable standards.

INTERNATIONAL ORGANIZATION FOR STANDARDIZATION (ISO)

As noted above, more than one standard can be written for a particular topic. Most of the overlapping standards come from industries within different countries. In 1946, the **International Organization for Standardization (ISO)**, was established. When this textbook was written, there were 162 members of ISO.[1] Per ISO, "[m]embership in ISO is open to national standards institutes most representative of standardization in their country (one member in each country)."[2] The ISO standards are voluntary standards developed by experts on an international level. ISO defines international standardization in the following way: "When the large majority of products or services in a particular business or industry sector conform to International Standards, a state of industry-wide standardization exists. The economic stakeholders concerned agree on specifications and criteria to be applied consistently in the classification of materials, in the manufacture and supply of products, in testing and analysis, in terminology and in the provision of services. In this way, International Standards provide a reference framework, or a common technological language, between suppliers and their customers. This facilitates trade and the transfer of technology."[3]

ISO WELDING SYMBOLS

In addition to the AWS welding symbol system described earlier in this text and defined in AWS A2.4:2007, Standard Symbols for Welding, Brazing and Nondestructive Examination, welders and fabricators should be aware of the international system defined in ISO 2553:1992 Welded, brazed and soldered joints—Symbolic representation on drawings. The base ISO welding symbol has an arrow, multiple reference line (solid and dashed, or solid only), and tail section when needed (see Figure 12-1). The arrow is used to point to the weld joint where the weld will be made. The reference line is where the information that shows the type of weld required is placed. The tail provides an area for reference information when needed.

---

[1] ISO International Organization for Standardization, *ISO Members,* http://www.iso.org/iso/about/iso_members.htm (accessed September 18, 2009).

[2] ISO International Organization for Standardization, *Who Can Join ISO,* http://www.iso.org/iso/about/discover-iso_who-can-join-iso.htm (accessed June 6, 2009).

[3] ISO International Organization for Standardization, *What "International Standardization" Means,* http://www.iso.org/iso/about/discover-iso_what-international-standardization-means.htm (accessed June 6, 2009).

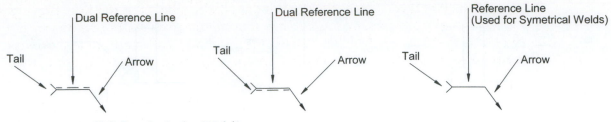

Figure 12-1   ISO Symbols for Welding

The dual reference line used in the ISO standard is one of the main differences between the AWS and ISO standards. The ISO dual reference line works in the following way (see Figure 12-2). When the weld symbol is placed on the solid reference line, the weld is to be placed on the arrow side of the weld joint. It does not matter whether the solid reference line is above or below the dashed reference line. When the weld symbol is placed on the dashed reference line, the weld is to be placed on the other side of the weld joint. When the weld is to be symmetrical (the same weld) on both sides of the weld joint, the multiple reference line is not used.

Just like the AWS arrow, the ISO arrow points at the weld joint. When only one of the plates is to be prepared, such as in the case of a single bevel or a J-groove, the arrow points to the plate to be prepared; however, it is not shown as a broken arrow like it is within the AWS system. The following pages contain figures that show the ISO and AWS symbols in arrow-side, other-side, and both-sides orientation, with examples of where the weld is to be placed. As you can see, there are minor differences, but with the exception of the dual reference line, many of the ISO symbols are the same or very similar to the AWS symbols.

Fillet welds for both ISO and AWS use the same weld symbol. As shown in Figure 12-3, the placement of the weld symbol is different depending on the use of the dual reference line.

Square groove welds for both ISO and AWS use the same weld symbol. As shown in Figure 12-4, the placement of the weld symbol is different depending on the use of the dual reference line.

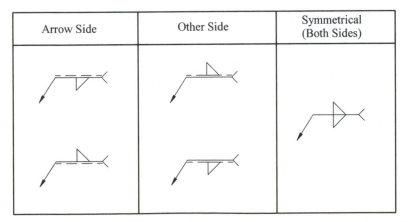

Arrow Side	Other Side	Symmetrical (Both Sides)

Figure 12-2   ISO Arrow Side/Other Side/Symmetrical Symbols

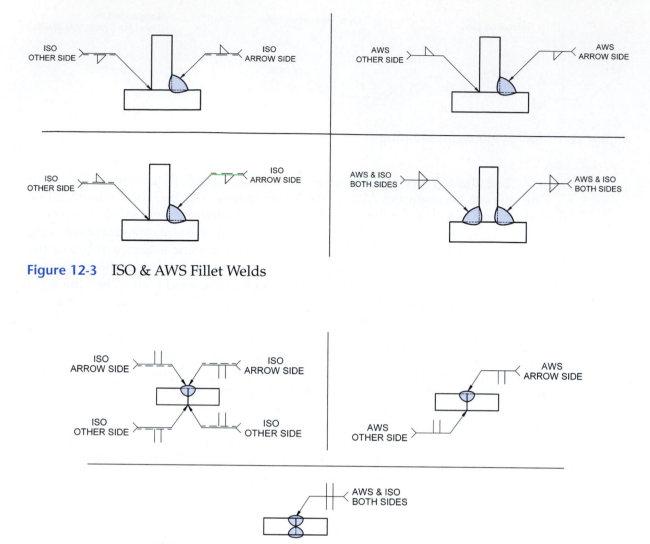

Figure 12-3   ISO & AWS Fillet Welds

Figure 12-4   ISO and AWS Square Groove Welds

Broad root face V-groove welds for ISO use a symbol not used by the AWS standard (see Figure 12-5). The AWS uses the standard V-groove symbol, along with the size and preparation measurements discussed in Chapter 10 to describe the depth of groove and the amount of root face required.

V-groove welds for both ISO and AWS use the same weld symbol. As shown in Figure 12-6, the placement of the weld symbol is different depending on the use of the dual reference line.

Bevel groove welds for both ISO and AWS use the same symbol. As shown in Figure 12-7, there are differences due to the use of the multiple reference line and the AWS use of the broken arrow to indicate the member to be beveled.

Broad root face bevel groove welds for ISO use a symbol not used by the AWS. The AWS uses a standard bevel groove symbol with the size of the root face given within the welding symbol. See Figure 12-8.

J-groove welds for both ISO and AWS use the same symbol. As shown in Figure 12-9, there are differences due to the use of the multiple reference line and the AWS use of the broken arrow to indicate the member to be beveled.

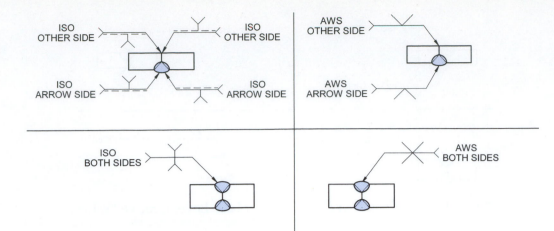

Figure 12-5 ISO and AWS V-Grooves with a Broad Root Face

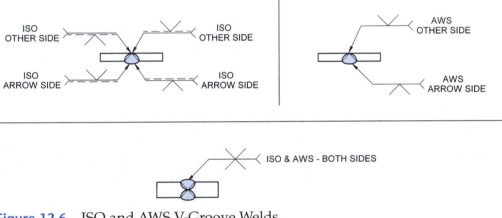

Figure 12-6 ISO and AWS V-Groove Welds

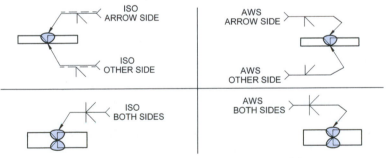

Figure 12-7 ISO and AWS Bevel Groove Welds

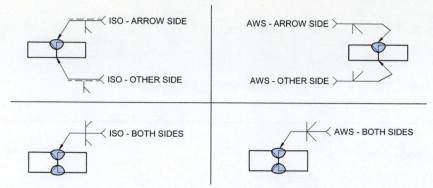

Figure 12-8 ISO and AWS Broad Root Face Bevel Groove Welds

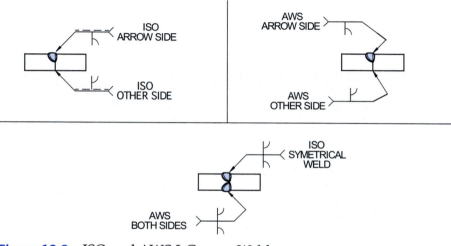

Figure 12-9 ISO and AWS J-Groove Welds

U-groove welds for both ISO and AWS use the same weld symbol. Figure 12-10 shows the placement of the weld symbol, which is different depending on the use of the dual reference line.

Flare bevel welds for both ISO and AWS use the same weld symbol. Figure 12-11 shows the placement of the weld symbol, which is different depending on the use of the dual reference line.

Flare-V welds for both ISO and AWS use the same weld symbol. As shown in Figure 12-12, the placement of the weld symbol is different depending on the use of the dual reference line.

Back and backing welds for both ISO and AWS use the same weld symbol. Figure 12-13 shows the difference in the placement of the weld symbol depending on the use of the dual reference line.

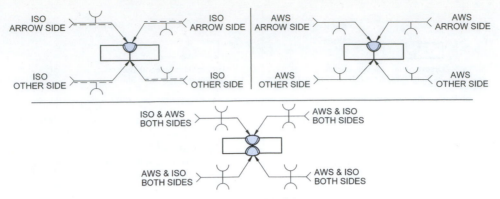

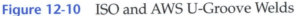

Figure 12-10   ISO and AWS U-Groove Welds

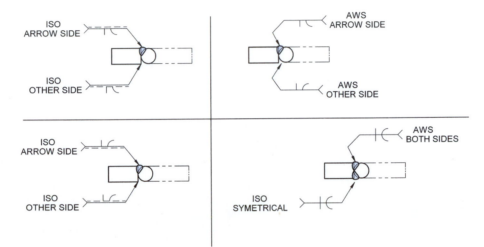

Figure 12-11   ISO and AWS Flare Bevel Welds

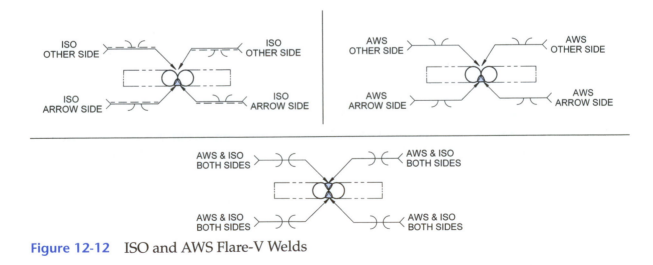

Figure 12-12   ISO and AWS Flare-V Welds

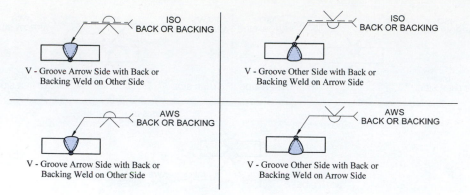

Figure 12-13   ISO Back and Backing Welds

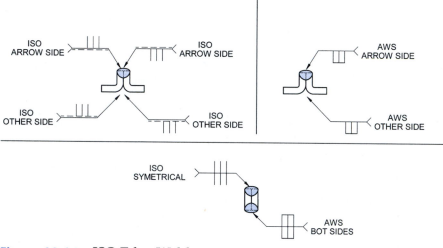

Figure 12-14   ISO Edge Welds

Edge welds for ISO, shown in Figure 12-14, are designated by a slightly different symbol than the one used in the AWS system. The other difference is the dual reference line.

Surfacing welds for both ISO and AWS use the same weld symbol. As shown in Figure 12-15, the placement of the weld symbol is different depending on the use of the dual reference line.

Spot welds for both ISO and AWS use the same weld symbol. As shown in Figure 12-16, the placement of the weld symbol is different depending on the use of the dual reference line.

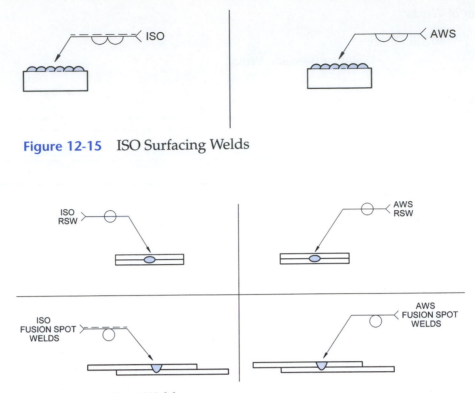

Figure 12-15　ISO Surfacing Welds

Figure 12-16　Spot Welds

ISO WELD DIMENSIONING AND ADDITIONAL WELD DESCRIPTION

Like the AWS system, the ISO weld size is placed to the left of the weld symbol and the length is placed to the right. The size of round welds (spot or plug) is given as the diameter. The size of seam welds is given as the width. When the length of the weld is not specified, it is considered to extend the entire length of the joint. The figures on the following pages show a few minor differences that should be understood when working with ISO style welding symbols.

INTERMITTENT WELDS

Intermittent welds are identified with the number of welds, the length of the welds, and the distance between the welds (see Figure 12-17). *Note:* This is different from the AWS system, where the number of welds is not given, and the pitch (center-to-center distance) of the welds is given, not the distance between the welds.

FILLET WELDS

The size of a fillet weld described by the ISO system is given by one of three different measurements, as shown in Figure 12-18. An s, a or z is placed in front of the numerical size on the welding symbol to indicate which size dimension is used.

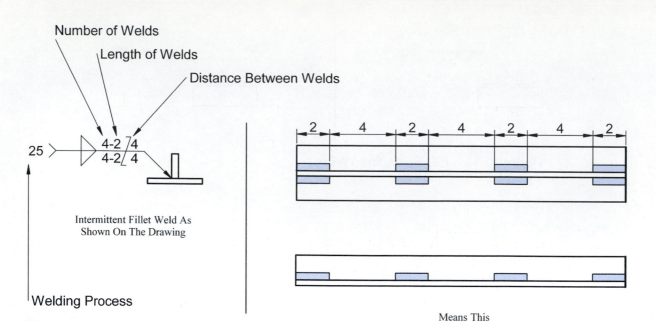

Figure 12-17   Method of Showing ISO Intermittent Welds

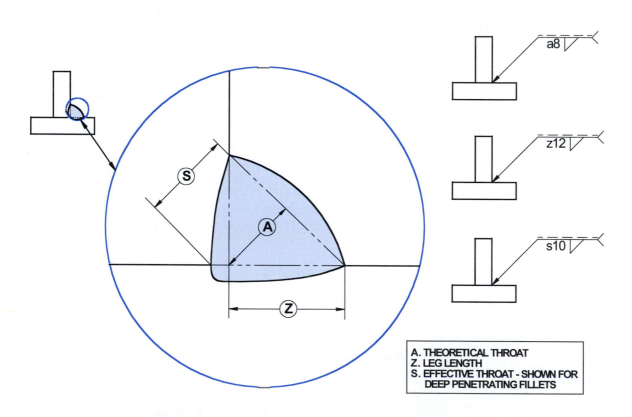

Figure 12-18   Different Methods for Showing Fillet Weld Size in the ISO System

BUTT WELDS

The size of AWS-defined groove welds, called butt welds in the ISO system, is defined in the ISO system as the distance from the surface of the base metal to the deepest penetration point, without being larger than the thickness of the metal (it doesn't include any face or root reinforcement). See Figure 12-19.

ISO WELDING SYMBOL COMPLIMENTARY INDICATIONS

ISO defines peripheral welds, site welds, process identifiers, and reference information as complimentary indications. Weld all around welds, called peripheral welds in the ISO system, are designated as such by a circle placed around the intersection between the arrow and the reference line. A flag placed at the intersection between the arrow and the reference line designates site welds (field welds). The required welding process for ISO welding symbols is listed in the tail section of the ISO welding symbol by a number that correlates back to a specific welding process defined in the standard ISO 4063. Other information and references can also be given in the tail section as needed; see Figure 12-20.

ISO SUPPLEMENTARY SYMBOLS

The ISO-defined supplementary symbols include weld contour symbols for flat/flush; convex and concave, which are the same as the AWS symbols; an additional symbol that is used when the toes of the weld are to be blended smoothly; and backing strip symbols for both permanent and removable backing. Examples of each are shown in Figures 12-21 and 12-22.

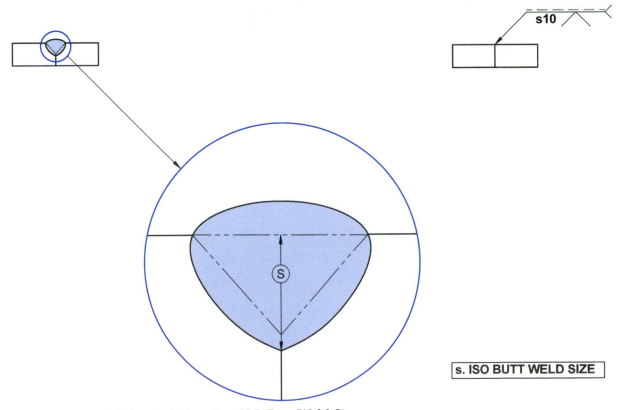

s. ISO BUTT WELD SIZE

Figure 12-19 Method of Showing ISO Butt Weld Size

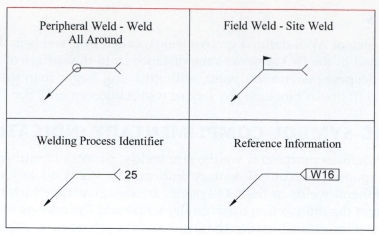

Peripheral Weld - Weld All Around	Field Weld - Site Weld
Welding Process Identifier 25	Reference Information W16

Figure 12-20   Methods of Showing Additional Weld Information

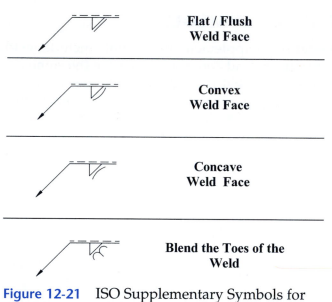

Flat / Flush
Weld Face

Convex
Weld Face

Concave
Weld Face

Blend the Toes of the
Weld

Figure 12-21   ISO Supplementary Symbols for Weld Contour with Examples

Permanent Backing Strip   M   **Removable Backing Strip**   MR

Figure 12-22   ISO Backing Strip Symbols

Practice Exercise 1

1. What ASME standard covers dimensioning and tolerancing? _____

2. Name two standards that cover how welds are to be represented on engineering drawings.

   A. _____

   B. _____

3. How is the arrow-side/other-side significance shown on an ISO welding symbol?

   _____

   _____

Describe completely all of the information contained in the following welding symbols.

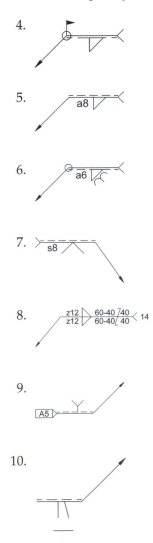

4.

5.

6.

7.

8.

9.

10.

CHAPTER 12

Math Supplement

Angles

An angle, like the one shown in Figure 12-23, is formed when two rays extend out from a single point called a vertex. All weld joints are put together at some angle. For instance, two plates that make up a standard T-joint are placed at a 90-degree angle to each other, whereas two plates that make up a standard butt joint are considered to be at a 180-degree angle to each other. Angular measurements typically shown on weldment drawings use the degree as the unit of measurement. The degree is an angular measurement of 1/360th of 1 complete revolution around a circle. Figure 12-24 shows a circle with 360 graduations, each representing 1 degree.

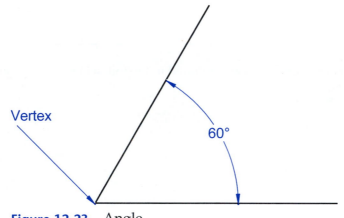

Figure 12-23 Angle

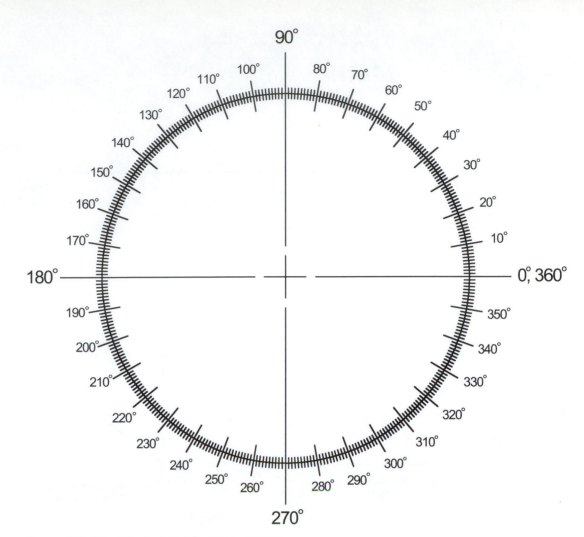

Figure 12-24 Circle Divided Into 360 Degrees

As shown in Figure 12-25, angles that are greater than 90 degrees are called obtuse angles, those that are exactly 90 degrees are called right angles, and those that are less than 90 degrees are called acute angles. A straight line is a 180-degree angle; it is called a straight angle (see Figure 12-26). Two angles that add together to equal 180 degrees are supplementary angles. Two angles that add together to equal 90 degrees are complementary angles (see Figure 12-27).

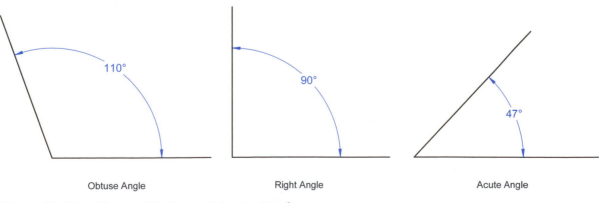

Figure 12-25 Obtuse, Right, and Acute Angles

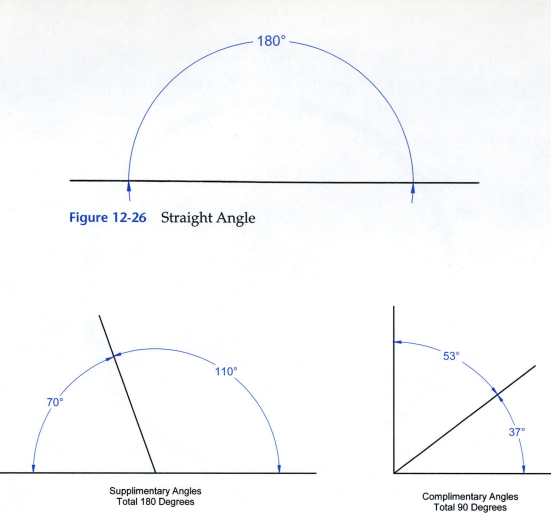

Figure 12-26   Straight Angle

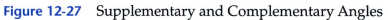

Supplimentary Angles
Total 180 Degrees

Complimentary Angles
Total 90 Degrees

Figure 12-27   Supplementary and Complementary Angles

Math Supplement Practice Exercise 1

1. Define point A. _____

2. What kind of angle is B? _____

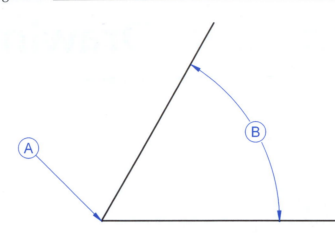

3. Are angles C and D known as supplementary or complementary angles? _____

4. Is angle E a right angle or a straight angle? _____

5. Is angle F a right angle or a straight angle? _____

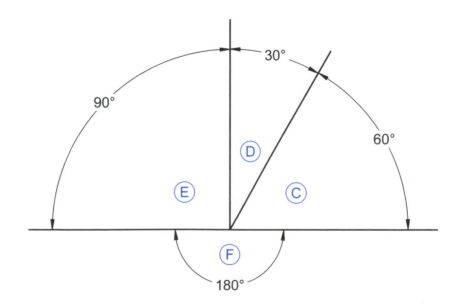

Additional Drawing Concepts

OBJECTIVE

- Understand a basic overview of some additional concepts that may be on drawings with which the welder or fabricator may work

KEY TERMS

Nondestructive examination (NDE)

Brinell

Knoop

Rockwell

Rockwell superficial

Heat treatment

Post-weld heat treatment (PWHT)

Stress relief

Annealing

Geometric dimensioning and tolerancing (GD&T)

OVERVIEW

Many designs that require welding and fabrication also include requirements that need additional symbols, notes, dimensioning, and tolerancing practices. Some of the common requirements are for inspection and testing of welds, material hardness, heat treating, machining plating/coating, and painting. It is important for the welder/fabricator to have a basic understanding of these additional principles and practices.

NONDESTRUCTIVE EXAMINATION (NDE)

Non-destructive examination (NDE) is any inspection or testing method that does not destroy the object being examined. Just like designs and drawings require welds to be made, they also require examination and testing of the welds so that their quality and integrity can be verified. A system similar to the welding symbol is used to communicate the type, location, and quantity of inspection and testing required. Figure 13-1 shows the standard American Welding Society NDE symbol.

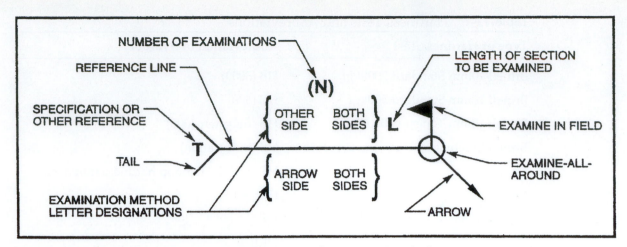

Figure 13-1 Standard Locations of Elements of an NDE Symbol *(AWS A2.4:2007, Figure 51 reproduced and adapted with permission from the American Welding Society (AWS), Miami, FL.)*

The basic inspection symbol consists of an arrow, reference line, and tail section when needed. The **arrow** is used to point to the weld where the inspection is required, the **reference line** is used to place information that shows the type of inspection required at the weld, and the **tail** provides an area for reference information. The basic examination symbols that are placed on the reference line are shown below:

RT = radiographic testing
PT = penetrant testing
VT = visual testing
UT = ultrasonic testing
MT = magnetic particle testing
AET = acoustic emission testing

HARDNESS REQUIREMENTS

For many applications, materials or specific areas of materials must meet hardness requirements. When this is the case, the hardness requirement and location are shown on the drawing. The welder/fabricator reading the print should be able to recognize the requirement shown and ensure that the correct material is used and the proper processing requirements are followed. The hardness for the material is shown on the drawing as a number and the scale the number is based on. For example, 55 HRC indicates a hardness number of 55 based on the Rockwell C scale. The higher the number within a specific scale, the harder the material. Typical methods for testing hardness and arriving at a particular hardness number are based on making an indentation in the metal and then measuring the indentation. The smaller the indentation in the material for a given testing method, the harder the metal is and the higher the hardness number will be for that particular scale. The names of the different types of hardness numbers are based on the particular type of testing used. Table 13-1

Table 13-1 Hardness Tests and Abbreviations

Type of Hardness Test	Abbreviation or Symbol
Brinell 10mm Standard 3000kgf	HB (3000)
Brinell 10mm Standard 500kgf	HB (500)
Brinell 10mm Tungsten 3000kgf	HB (tungsten 3000)
Brinell Indentation	HB (indentation)
Knoop	HK or KHN (Knoop hardness number)
Mohs	HM
Rockwell A	HRA
Rockwell B	HRB
Rockwell C	HRC
Rockwell D	HRD
Rockwell F	HRF
Rockwell Superficial 15N	HR-15N
Rockwell Superficial 15T	HR-15T
Rockwell Superficial 30N	HR-30N
Rockwell Superficial 30T	HR-30T
Rockwell Superficial 45N	HR-45N
Rockwell Superficial 45T	HR-45T
Shore Scleroscope	HS
Vickers	HV

shows different types of hardness tests with the abbreviation that may be used on a drawing to represent each test.

Additional information regarding hardening requirements may be referenced in a note, standard, or specification. The location of the hardness requirement may be identified through the use of standard dimensioning methods, notes, or a combination thereof. Hardness numbers from a test type often need to be converted to another test type. Table 13-2 is a hardness conversion table from ASTM E140-07. This particular table is specific to hardness values of a particular range of hardness (Rockwell C hardness range) on a specific type of material (nonaustenitic steel). The table shows how the different hardness numbers from different testing methods correlate to each other. There are many tables like this one that cover other materials and hardness ranges.

Table 13-2   Hardness Comparison Table

⟨ASTM⟩ E 140 – 07

TABLE 1 Approximate Hardness Conversion Numbers for Non-Austenitic Steels (Rockwell C Hardness Range)[A, B]

Rockwell C Hardness Number 150 kgf (HRC)	Vickers Hardness Number (HV)	Brinell Hardness Number[C]		Knoop Hardness, Number 500-gf and Over (HK)	Rockwell Hardness Number		Rockwell Superficial Hardness Number			Sclero-scope Hard-ness Number[D]	Rockwell C Hardness Number 150 kgf (HRC)
		10-mm Standard Ball, 3000-kgf (HBS)	10-mm Carbide Ball, 3000-kgf (HBW)		A Scale, 60-kgf (HRA)	D Scale, 100-kgf (HRD)	15-N Scale, 15-kgf (HR 15-N)	30-N Scale, 30-kgf (HR 30-N)	45-N Scale, 45-kgf (HR 45-N)		
68	940	...	...	920	85.6	76.9	93.2	84.4	75.4	97.3	68
67	900	...	...	895	85.0	76.1	92.9	83.6	74.2	95.0	67
66	865	...	...	870	84.5	75.4	92.5	82.8	73.3	92.7	66
65	832	...	(739)	846	83.9	74.5	92.2	81.9	72.0	90.6	65
64	800	...	(722)	822	83.4	73.8	91.8	81.1	71.0	88.5	64
63	772	...	(705)	799	82.8	73.0	91.4	80.1	69.9	86.5	63
62	746	...	(688)	776	82.3	72.2	91.1	79.3	68.8	84.5	62
61	720	...	(670)	754	81.8	71.5	90.7	78.4	67.7	82.6	61
60	697	...	(654)	732	81.2	70.7	90.2	77.5	66.6	80.8	60
59	674	...	634	710	80.7	69.9	89.8	76.6	65.5	79.0	59
58	653	...	615	690	80.1	69.2	89.3	75.7	64.3	77.3	58
57	633	...	595	670	79.6	68.5	88.9	74.8	63.2	75.6	57
56	613	...	577	650	79.0	67.7	88.3	73.9	62.0	74.0	56
55	595	...	560	630	78.5	66.9	87.9	73.0	60.9	72.4	55
54	577	...	543	612	78.0	66.1	87.4	72.0	59.8	70.9	54
53	560	...	525	594	77.4	65.4	86.9	71.2	58.6	69.4	53
52	544	(500)	512	576	76.8	64.6	86.4	70.2	57.4	67.9	52
51	528	(487)	496	558	76.3	63.8	85.9	69.4	56.1	66.5	51
50	513	(475)	481	542	75.9	63.1	85.5	68.5	55.0	65.1	50
49	498	(464)	469	526	75.2	62.1	85.0	67.6	53.8	63.7	49
48	484	451	455	510	74.7	61.4	84.5	66.7	52.5	62.4	48
47	471	442	443	495	74.1	60.8	83.9	65.8	51.4	61.1	47
46	458	432	432	480	73.6	60.0	83.5	64.8	50.3	59.8	46
45	446	421	421	466	73.1	59.2	83.0	64.0	49.0	58.5	45
44	434	409	409	452	72.5	58.5	82.5	63.1	47.8	57.3	44
43	423	400	400	438	72.0	57.7	82.0	62.2	46.7	56.1	43
42	412	390	390	426	71.5	56.9	81.5	61.3	45.5	54.9	42
41	402	381	381	414	70.9	56.2	80.9	60.4	44.3	53.7	41
40	392	371	371	402	70.4	55.4	80.4	59.5	43.1	52.6	40
39	382	362	362	391	69.9	54.6	79.9	58.6	41.9	51.5	39
38	372	353	353	380	69.4	53.8	79.4	57.7	40.8	50.4	38
37	363	344	344	370	68.9	53.1	78.8	56.8	39.6	49.3	37
36	354	336	336	360	68.4	52.3	78.3	55.9	38.4	48.2	36
35	345	327	327	351	67.9	51.5	77.7	55.0	37.2	47.1	35
34	336	319	319	342	67.4	50.8	77.2	54.2	36.1	46.1	34
33	327	311	311	334	66.8	50.0	76.6	53.3	34.9	45.1	33
32	318	301	301	326	66.3	49.2	76.1	52.1	33.7	44.1	32
31	310	294	294	318	65.8	48.4	75.6	51.3	32.5	43.1	31
30	302	286	286	311	65.3	47.7	75.0	50.4	31.3	42.2	30
29	294	279	279	304	64.8	47.0	74.5	49.5	30.1	41.3	29
28	286	271	271	297	64.3	46.1	73.9	48.6	28.9	40.4	28
27	279	264	264	290	63.8	45.2	73.3	47.7	27.8	39.5	27
26	272	258	258	284	63.3	44.6	72.8	46.8	26.7	38.7	26
25	266	253	253	278	62.8	43.8	72.2	45.9	25.5	37.8	25
24	260	247	247	272	62.4	43.1	71.6	45.0	24.3	37.0	24
23	254	243	243	266	62.0	42.1	71.0	44.0	23.1	36.3	23
22	248	237	237	261	61.5	41.6	70.5	43.2	22.0	35.5	22
21	243	231	231	256	61.0	40.9	69.9	42.3	20.7	34.8	21
20	238	226	226	251	60.5	40.1	69.4	41.5	19.6	34.2	20

[A] In the table headings, *force* refers to total test forces.

[B] Appendix X1 contains equations converting determined hardness scale numbers to Rockwell C hardness numbers for non-austenitic steels. Refer to 1.11 before using conversion equations.

[C] The Brinell hardness numbers in parentheses are outside the range recommended for Brinell hardness testing in 8.1 of Test Method E 10.

[D] These Scleroscope hardness conversions are based on Vickers—Scleroscope hardness relationships developed from Vickers hardness data provided by the National Bureau of Standards for 13 steel reference blocks, Scleroscope hardness values obtained on these blocks by the Shore Instrument and Mfg. Co., Inc., the Roll Manufacturers Institute, and members of this institute, and also on hardness conversions previously published by the American Society for Metals and the Roll Manufacturers Institute.

Source: Reprinted, with permission from ASTM E140-07 Standard Hardness Conversion Tables metals relationship among Brinell hardness, Vickers hardness, Rockwell hardness, Superficial hardness, Knoop hardness, Copyright ASTM International (WWW.ASTM.ORG).

HEAT-TREATING REQUIREMENTS

For many applications, metals and fabrications require heating operations prior, during, or after welding. The method in which the heat is applied, the length of time, and the particular time at which the heat is applied are all factors that must be controlled to achieve the desired outcome. When heat treatment is required, the drawing should include the requirements on the drawing, or it should include a note referencing other documents or specifications that give the specific requirements of the treatment. A few basic heat treatment terms are shown below with their definitions. Heat treatment of metals is a very broad and detailed subject covering many different processes and methods, and it is beyond the scope of this textbook. Welders or fabricators who need a greater understanding of heat treatment will need coursework and study beyond the material covered in this text.

preheat: Applying heat to the base metal prior to welding/joining or cutting processes.

postheat: Applying heat to the welded part after welding. Also known as post-weld heat treatment (PWHT).

interpass temperature: The temperature of the base metal between weld passes.

annealing: Any one of a variety of methods where steel is heated to a certain temperature, held at the temperature, and then cooled at a certain rate in order to achieve desired properties of the metal.

stress relief: A method of relieving stresses in a weldment by uniform heating and cooling.

SURFACE TEXTURE SYMBOLS

When the surface of an object is required to be a certain texture, a surface texture symbol is used. Figure 13-2 shows the surface texture symbols and their construction.

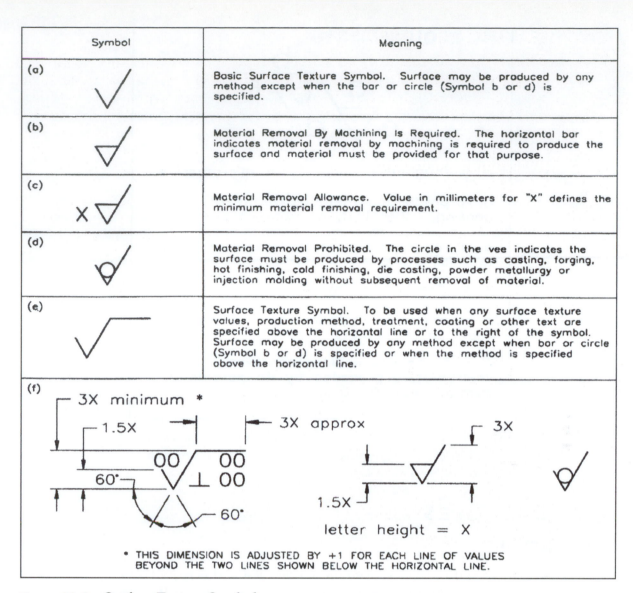

	Symbol	Meaning
(a)		Basic Surface Texture Symbol. Surface may be produced by any method except when the bar or circle (Symbol b or d) is specified.
(b)		Material Removal By Machining Is Required. The horizontal bar indicates material removal by machining is required to produce the surface and material must be provided for that purpose.
(c)		Material Removal Allowance. Value in millimeters for "X" defines the minimum material removal requirement.
(d)		Material Removal Prohibited. The circle in the vee indicates the surface must be produced by processes such as casting, forging, hot finishing, cold finishing, die casting, powder metallurgy or injection molding without subsequent removal of material.
(e)		Surface Texture Symbol. To be used when any surface texture values, production method, treatment, coating or other text are specified above the horizontal line or to the right of the symbol. Surface may be produced by any method except when bar or circle (Symbol b or d) is specified or when the method is specified above the horizontal line.
(f)		

Figure 13-2 Surface Texture Symbols *(Reprinted from ASME Y14.36M 1996 by permission of the American Society of Mechanical Engineers. All rights reserved.)*

SURFACE TEXTURE SYMBOL VALUES

Letters and numbers are used with the surface texture symbols to provide additional information regarding the surface that is to be textured. Figure 13-3 shows location and description details for this additional information.

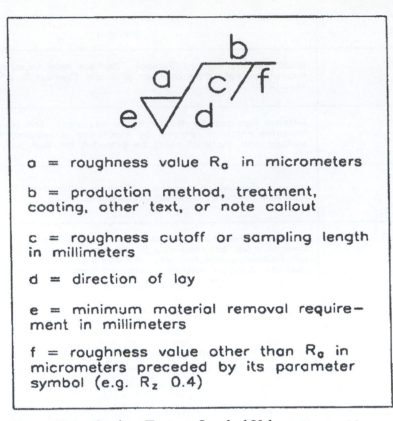

a = roughness value R_a in micrometers

b = production method, treatment, coating, other text, or note callout

c = roughness cutoff or sampling length in millimeters

d = direction of lay

e = minimum material removal requirement in millimeters

f = roughness value other than R_a in micrometers preceded by its parameter symbol (e.g. R_z 0.4)

Figure 13-3 Surface Texture Symbol Values *(Reprinted from ASME Y14.36M 1996 by permission of the American Society of Mechanical Engineers. All rights reserved.)*

LAY SYMBOLS

When the surface texture on the surface of a part is required to lay in a certain direction, this information is noted on the drawing by the placement of a lay symbol. Each of the lay symbols, along with the corresponding meaning, is shown in Figure 13-4. Although the welder/fabricator may not be the one making the pattern on the material, it may fall upon him or her to ensure that the material is properly placed according to the direction of the tool marks during assembly or welding operations.

PIPE DRAWINGS

Piping designs can be represented many different ways. The method of representation typically depends on the complexity of the application, the requirements of the particular job, contract documents, and company practices. They may be drawn in 3D type pictorial views or in orthographic views. Two different methods are shown in Figures 13-5 and 13-6. Figure 13-5 shows a two-line view using a standard welding symbol to indicate the type of weld required. Notice that the pipe shown in Figure 13-5 does not include hidden lines to indicate the inside of the pipe. This is common on a pipe drawing and is done to simplify the drawing.

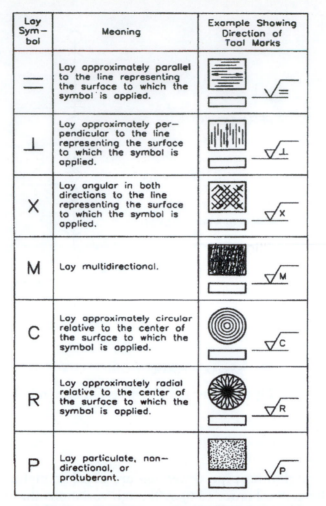

Lay Symbol	Meaning	Example Showing Direction of Tool Marks
═	Lay approximately parallel to the line representing the surface to which the symbol is applied.	
⊥	Lay approximately perpendicular to the line representing the surface to which the symbol is applied.	
X	Lay angular in both directions to the line representing the surface to which the symbol is applied.	
M	Lay multidirectional.	
C	Lay approximately circular relative to the center of the surface to which the symbol is applied.	
R	Lay approximately radial relative to the center of the surface to which the symbol is applied.	
P	Lay particulate, non-directional, or protuberant.	

Figure 13-4 Lay Symbols *(Reprinted from ASME Y14.36M 1996 by permission of the American Society of Mechanical Engineers. All rights reserved.)*

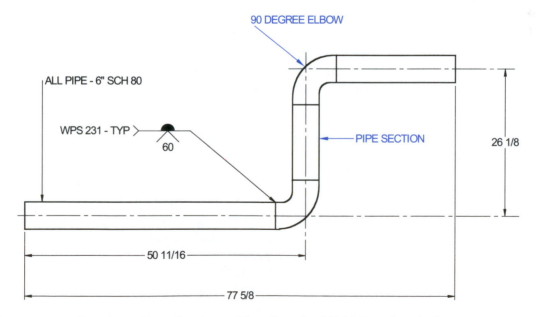

Figure 13-5 Two Line Pipe Section with a Standard Welding Symbol

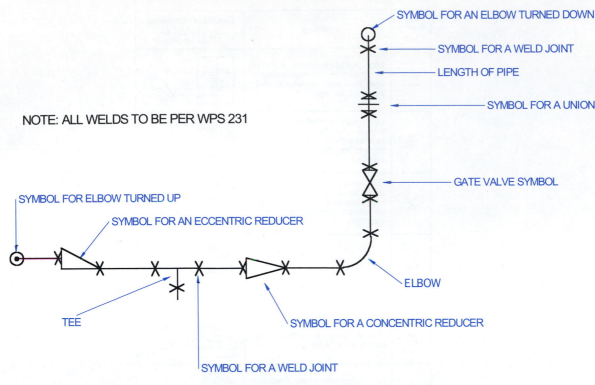

NOTE: ALL WELDS TO BE PER WPS 231

SYMBOL FOR AN ELBOW TURNED DOWN

SYMBOL FOR A WELD JOINT

LENGTH OF PIPE

SYMBOL FOR A UNION

GATE VALVE SYMBOL

ELBOW

SYMBOL FOR ELBOW TURNED UP

SYMBOL FOR AN ECCENTRIC REDUCER

TEE

SYMBOL FOR A CONCENTRIC REDUCER

SYMBOL FOR A WELD JOINT

Figure 13-6   Single Line Pipe Layout with Various Other Symbols

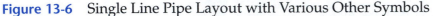

Single-line pipe drawings are another method of representing piping systems (see Figure 13-6). Although variations throughout industry may exist, an X placed either right-side-up or on its side on the single-line layout typically indicates a weld joint. Other symbols are used at pipe joints where the joints are to be flanged, screwed, soldered, etc. There can be many different components to a piping system, including straight pipe, 90-degree elbow, 45-degree elbow, union, Tee, valve, concentric reducer, and eccentric reducer. Many standard and nonstandard symbols have been used in industry to identify these components on drawings; a few are shown in Figure 13-6. When working with pipe drawings, welders should become familiar with the symbols and drawing conventions used for the particular set of prints being read.

GEOMETRIC DIMENSIONING AND TOLERANCING (GD&T)

Geometric dimensioning and tolerancing (GD&T) is a system used to describe the geometry (shape) of a part and its allowable variation. When the system is properly used and understood, it is very effective in defining geometric requirements and relationships, reducing costs, and providing required information for inspection. Company standards and customer and industry requirements normally dictate whether or not GD&T is used. Weldment drawings that include secondary operations such as machining are more likely to include GD&T than those that do not. The welder/fabricator should be able to recognize the GD&T system and symbols so that when working with drawings that contain GD&T, he or she can communicate with supervision, machining department personnel, or engineering personnel as required to ensure that the welding operations can be performed properly and do not adversely affect the GD&T requirements. Welders or fabricators who need a working understanding of GD&T will need coursework and study beyond the material offered in this text, which includes only a basic descriptive overview of GD&T. Use and interpretation of GD&T without proper training should be avoided. The American Society of Mechanical Engineers (ASME) and others offer classes covering GD&T. Both the ASME and the International Standards Organization (ISO) have published standards that describe GD&T in great detail. The following information is based on the American Society of Mechanical Engineers Standard ASME Y14.5–2009.

Datum

A **datum** is "a theoretically exact point, axis, line, plane, or combination thereof derived from the theoretical datum feature simulator."[1] It is identified by a capital letter within a square or rectangle frame that is attached to a leader that extends to the feature. See Figure 13-7.

Features and the Feature Control Frame

Surfaces, holes, slots, and pins are all features. A feature control frame is a rectangular frame with sections that provide the specific information that pertains to a feature. The number of sections contained can vary depending on the information required. The information is placed in a specific order so that it can be interpreted properly. Figure 13-8 shows an example of a feature control frame with a basic explanation of its contents.

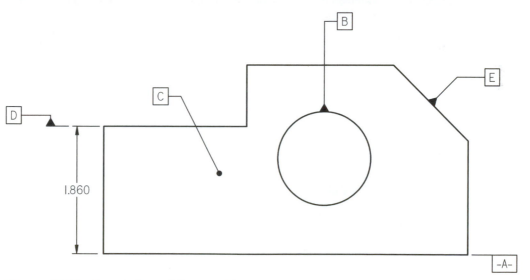

Figure 13-7 Examples of Datum Placement

---

[1]American Society of Mechanical Engineers, ASME Y14.5-2009, 3.

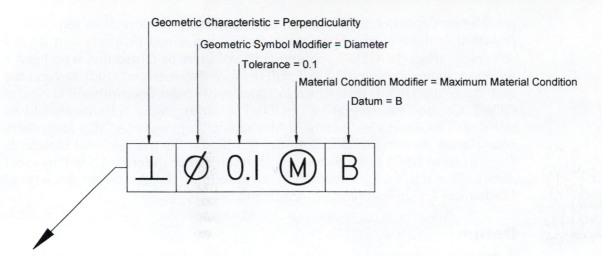

Figure 13-8   Sample Feature Control Frame

Figure 13-9 shows the different symbols for the geometric characteristics that can be defined using GD&T. Flatness, straightness, perpendicularity, angularity, and parallelism are the geometric characteristics that are typically used in welding operations when GT&T is applied.

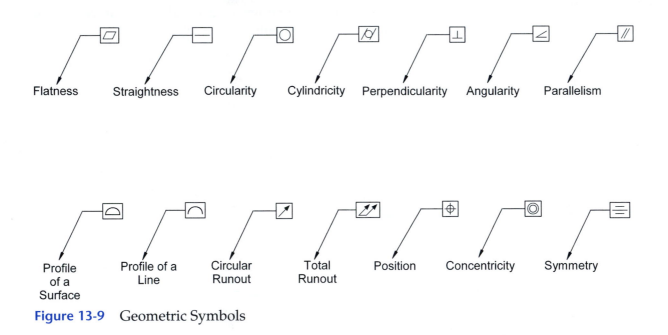

Figure 13-9   Geometric Symbols

When welding is required and the tolerances on applicable geometric characteristics are held tight, the welding is typically done prior to a machining process. In such cases, the welder uses as much precision and thermal distortion reduction as is practical for the welding process so that, when the machining process is carried out, it can be done effectively. In such cases, the final product after machining is typically what has to meet the GD&T requirements. When welding is the final process and GD&T is used, the tolerances applied should be large enough to accommodate enough variation so that the requirements can be met even with the thermal distortion caused by the welding process.

CHAPTER 13

Practice Exercise 1

Answer the following questions.

1. What does NDE stand for? _____

Describe the testing required for the following symbols.

2. _____

3. _____

4. _____

5. What does HRC stand for? _____

6. What does HV stand for? _____

7. Is 45 HRC harder than 60 HRC? _____

8. Is 50 HRC harder than 302 HV? _____

9. What is the approximate HRC equivalent to 720 HV? _____

10. What does PWHT stand for? _____

For questions 11 through 14, describe the surface symbols completely.

11. _____

12. _____

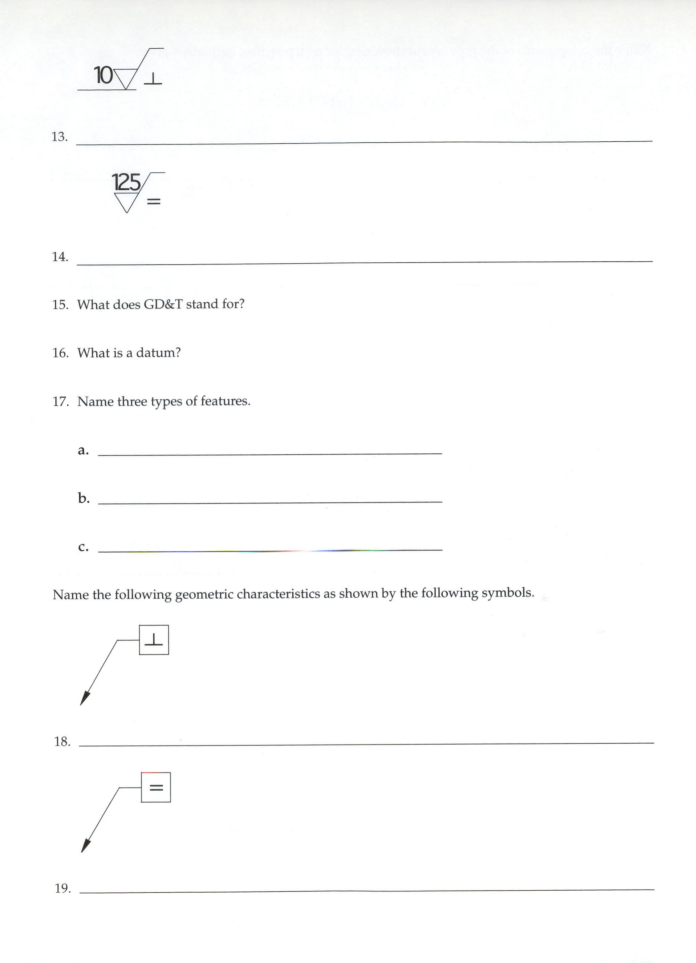

13. _____

14. _____

15. What does GD&T stand for?

16. What is a datum?

17. Name three types of features.

   a. _____

   b. _____

   c. _____

Name the following geometric characteristics as shown by the following symbols.

18. _____

19. _____

Name the components of the pipe layout shown, based on the symbol definitions given in Figure 13-6.

20. _____

21. _____

22. _____

23. _____

24. _____

25. _____

26. _____

27. _____

28. _____

29. _____

30. _____

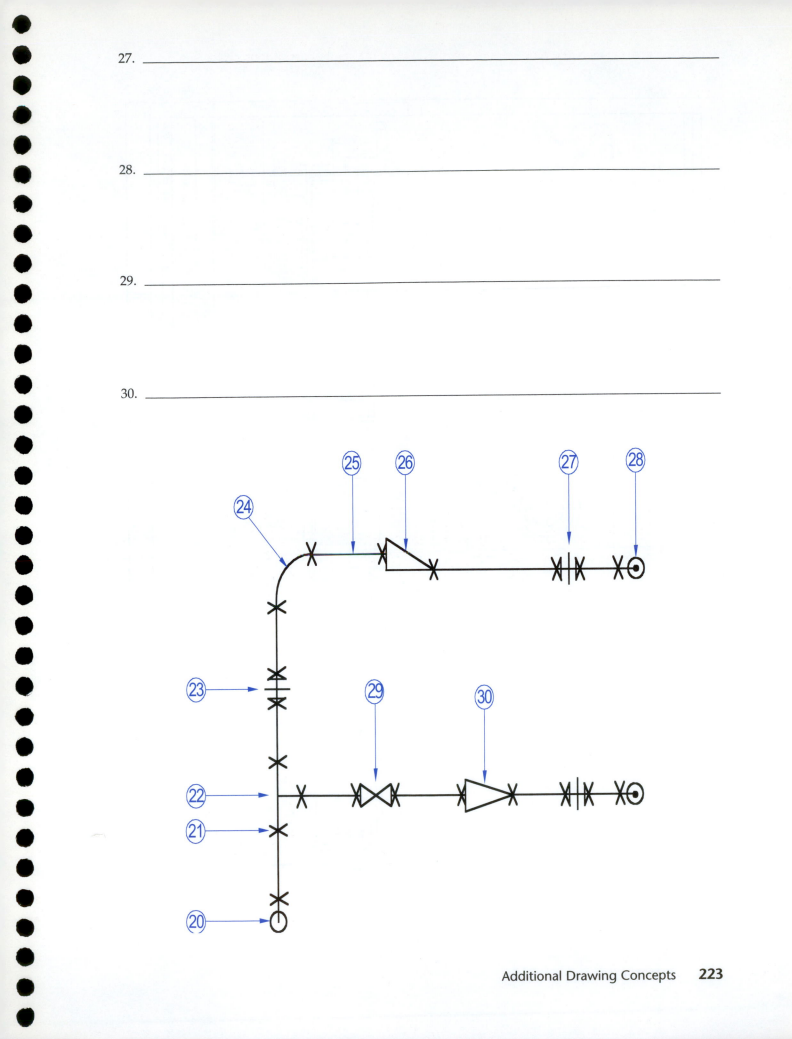

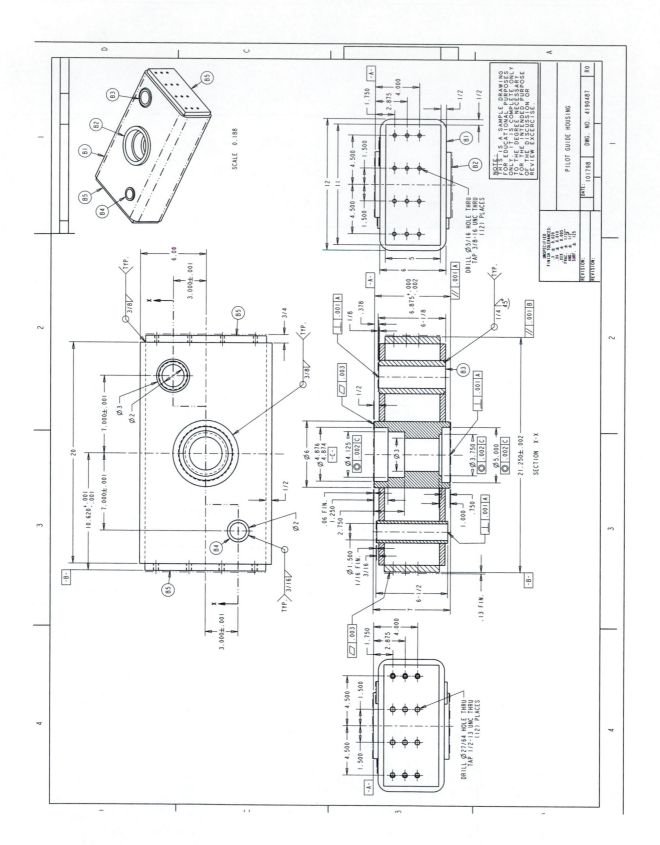

CHAPTER 13

Practice Exercise 2

Refer to Drawing Number 4190487 R0 on page 224 and answer the following questions.

1. What type of section view is shown?

   _____

2. How is B2 attached to B1? _____

   _____

3. What size is the bored hole in B4?

   _____

4. How many tapped holes are in the machined housing? _____

5. What size are the tapped holes? _____

   _____

6. How many geometric control frames are shown? _____

7. How many different geometric characteristics are shown? _____

8. Name the geometric characteristics shown on this drawing. _____

   _____

9. What is the largest diameter bored pocket in B2? _____

10. What is the smallest diameter bored pocket in B2? _____

11. What is the non-machined inside diameter (ID) of B2? _____

12. What is the finish allowance shown at datum B? _____

13. What is the length of B4? _____

14. What is the overall length of the assembly before finishing? _____

15. What is the largest tolerance shown on this drawing? _____

16. What does this tolerance apply to?

    _____

17. How is B5 attached to B1? _____

    _____

18. What are the dimensions of B5? _____

    _____

19. To what datum is the center line of the bore in B4 to be perpendicular? _____

20. What is the contour of the weld that attaches B3 to B1? _____

CHAPTER 13

Math Supplement

Degree Conversions

As described in the Math Supplement in Chapter 12, the degree is the unit typically used on weldment drawings to describe the size of an angle. Figure 13-10 shows one complete revolution around the circumference of a circle divided into 360 degrees.

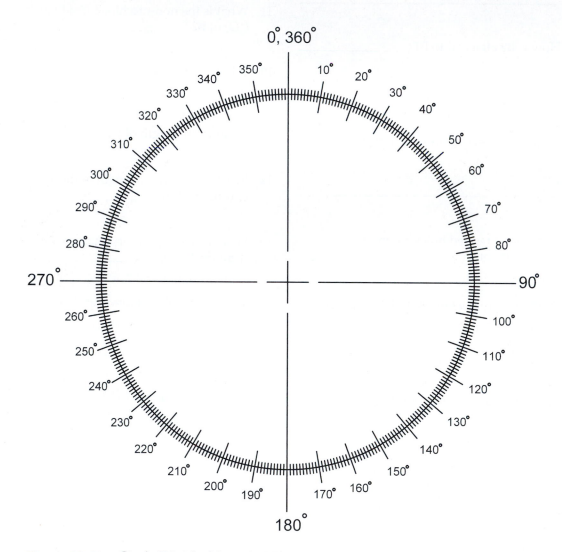

Figure 13-10   Circle Divided Into 360 Degrees

Each degree can be further divided into smaller increments consisting of common fractions, decimal fractions, or minutes and seconds.

- Examples of common fraction degree measurements are ½°, 90½°, 60¼°.
- Examples of decimal fraction degree measurements are .5°, 120.3°, 90.8°.

One **minute** is 1/60 of one degree. In other words, there are 60 minutes in 1 degree. The symbol used to designate minutes is the single hash mark: '. One **second** is 1/60 of 1 minute. In other words, there are 60 seconds in 1 minute. The symbol for seconds is dual hash mark:'.

- Examples of degree, minute, seconds measurements are 30°45', 60°30'27".

To convert minutes to a decimal fraction of a degree, divide the number of minutes by 60.

Example: 45' = 45/60 = .75°

To convert seconds to a decimal fraction of a degree, divide the number of seconds by 3600.

Example: 30" = 30/3600 = .00833°

To convert a number containing both minutes and seconds to a decimal fraction of a degree, convert each one individually and then add them together. See the conversion in the example below for 60°30'27".

Example: 60° + 30/60 + 27/3600 = 60° + .5° + .0075°

Answer: 60.50075°

Finding The Length Of An Arc

When a degree dimension is given for an item located on a circumference it can be necessary to find the linear distance of the arc so that the item can be properly located. When this is required, the following formula can be used.

$$\frac{C}{360} \cdot D = \text{Arc Length}$$

Where:

C = Circumference
D = Degrees shown on the dimension for the object being placed on the circumference

The following three steps describe how to use the above formula.

Step 1: Measure the circumference of the pipe, tank, cylinder, etc. to get the actual circumference. Note: The calculated value ($\pi \cdot D$) of the circumference can be used to achieve a calculated result, however the actual measured circumference should be used to obtain a result that reflects the actual size of the part.

Step 2: Divide the circumference by 360 to determine the unit length per degree

Step 3: Multiply the answer from step 2 by the degree measurement given on the drawing for the item being placed on the circumference.

Example: The centerline of a nozzle is to be place at 60 degrees from top dead center (TDC) on a tank that is supposed to have an O.D. of 58-1/2". How many inches from TDC is the centerline of the nozzle to be located?

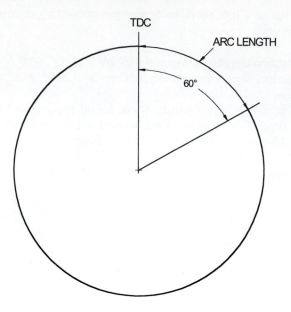

Measured Circumference = 183.5"

Step 1: Measure the circumference of the tank. For this example, assume it measures 183.5".

Step 2: Divide the measured circumference by 360 to get .5097 inches per degree.

Step 3: Multiply .5097 by 60 degrees to get 30.582", which is approximately 30-9/16"

Or as shown all together using the formula $\dfrac{C}{360} \cdot D = \text{Arc Length}$

$$\dfrac{183.5}{360} \cdot 60 = 30.582'' \approx 30\text{-}9/16''$$

Math Supplement Practice Exercise 1

Convert the following degree, minute, second measurements to degree decimal measurements.

1. 42°30′ _____

2. 105°15′ _____

3. 90°45′0″ _____

4. 62°20′15″ _____

5. The centerline of a nozzle is to be place at 18 degrees from top dead center (TDC) on a tank with a measured circumference of 280″. How many inches from TDC is the centerline of the nozzle to be located? _____

6. The centerline of a nozzle is to be place at 120 degrees from top dead center (TDC) on a tank with a measured circumference of 315″. How many inches from TDC is the centerline of the nozzle to be located? _____

Review Exercises

OBJECTIVE

- Use all of the knowledge gained in the first thirteen chapters for reading the prints and answering the questions in this chapter

NOTE

- Drawing Number 09022601 was drawn by the author. The remaining prints in this chapter were drawn by Kevin Schaaf, with collaboration from the author. The questions written for the drawings made by Kevin Schaaf were a collaborative effort by the author and Kevin Schaaf.

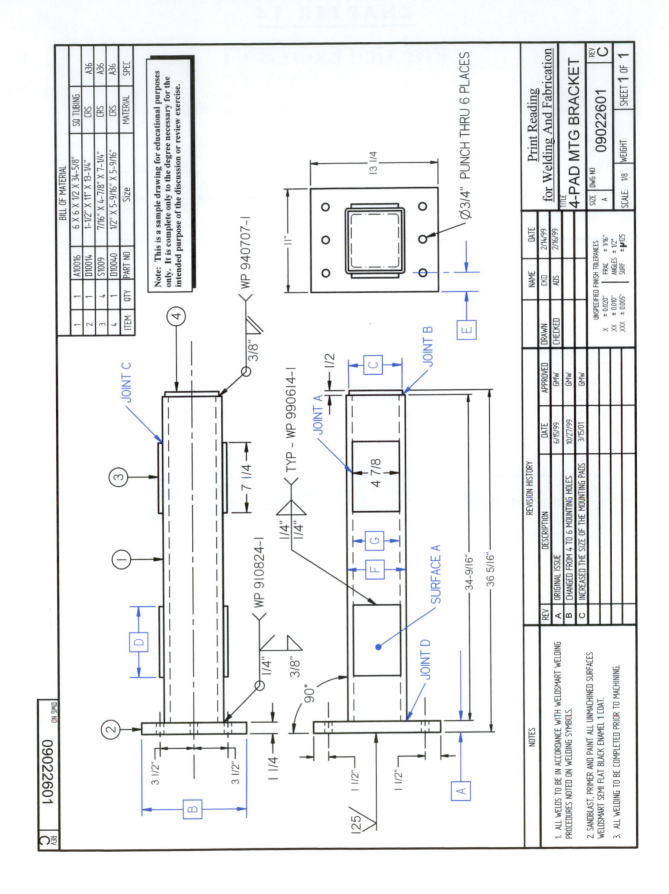

BILL OF MATERIAL

ITEM	QTY	PART NO	Size	MATERIAL	SPEC
1	1	A10016	6 X 6 X 1/2 X 34-5/8"	SQ TUBING	A36
2	1	D10014	1-1/2" X 11" X 13-1/4"	CRS	A36
3	4	S1009	7/16" X 4-7/8" X 7-1/4"	CRS	A36
4	1	D10040	1/2" X 5-9/16" X 5-9/16"	CRS	A36

Note: This is a sample drawing for educational purposes only. It is complete only to the degree necessary for the intended purpose of the discussion or review exercise.

JOINT C

WP 940707-1

3/8"

7 1/4

WP 910824-1

1/4" TYP – WP 990614-1

1/4"

3/8"

1/4"

90°

D

B

3 1/2"   3 1/2"

1 1/4"

13 1/4

11"

Ø3/4" PUNCH THRU 6 PLACES

E

JOINT A

JOINT B

1/2

C

4 7/8

G

F

SURFACE A

JOINT D

34-9/16"

36 5/16"

125

1 1/2"   1 1/2"

A

Print Reading
for Welding And Fabrication

TITLE		

4-PAD MTG BRACKET

SIZE	DWG NO	REV
A	09022601	C
SCALE: 1/8	WEIGHT:	SHEET 1 OF 1

	NAME	DATE
DRAWN		
CHECKED	CKD	2/14/99
	ADS	2/16/99

UNSPECIFIED FINISH TOLERANCES
X ± 0.020"   FRAC ± 1/16"
XX ± 0.010" ANGLES ± 1/2°
XXX ± 0.005" SURF ± 125

REVISION HISTORY

REV	DESCRIPTION	DATE	APPROVED
A	ORIGINAL ISSUE	6/15/99	GMW
B	CHANGED FROM 4 TO 6 MOUNTING HOLES	10/27/99	GMW
C	INCREASED THE SIZE OF THE MOUNTING PADS	3/15/01	GMW

NOTES
1. ALL WELDS TO BE IN ACCORDANCE WITH WELDSMART WELDING PROCEDURES NOTED ON WELDING SYMBOLS.
2. SANDBLAST, PRIMER AND PAINT ALL UNMACHINED SURFACES WELDSMART SEMI FLAT BLACK ENAMEL 1 COAT.
3. ALL WELDING TO BE COMPLETED PRIOR TO MACHINING.

DWG NO 09022601   REV C

CHAPTER 14
Practice Exercise 1

Refer to Drawing Number 09022601 REV C on page 231 and answer the following questions.

1. What is distance A? _____

2. What is distance B? _____

3. What is distance C? _____

4. What is distance D? _____

5. What is distance E? _____

6. What is distance F? _____

7. What is distance G? _____

8. Is there any requirement for the machining of surface A? _____

9. If this part were fabricated in July 1999, how many mounting holes would it have had?

   _____

10. What is the size of the mounting holes? _____

11. How much material is to be removed from Item 2 during the machining process?

    _____

12. How thick is Item 3? _____

13. Fully describe the weld, if one is required, at joint A. _____

    _____

    _____

14. Fully describe the weld, if one is required, at joint B. _____

    _____

    _____

15. Fully describe the weld, if one is required, at joint C. _____

    _____

    _____

16. Fully describe the weld, if one is required, at joint D.

    _____

    _____

    _____

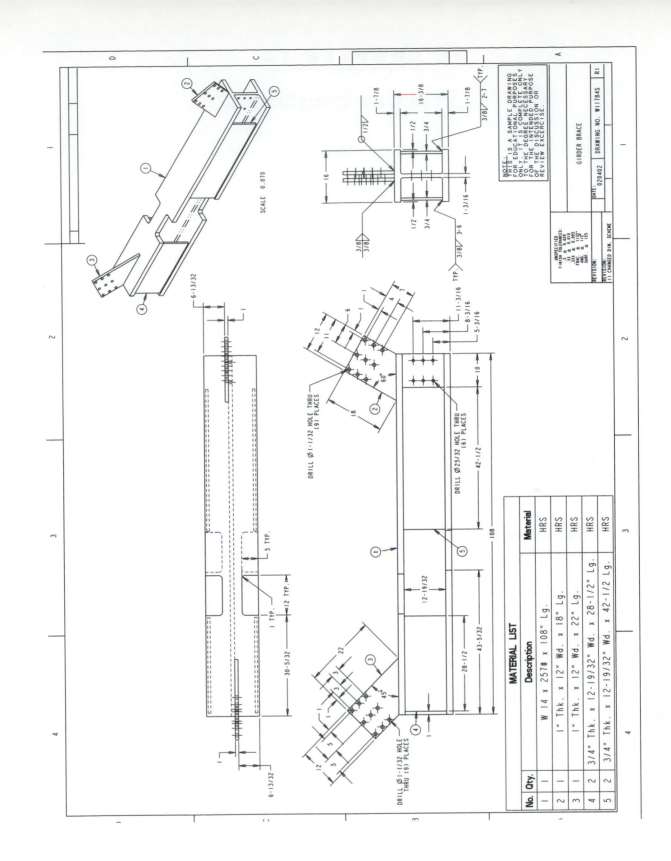

SCALE 0.070

MATERIAL LIST

No.	Qty.	Description	Material
1	1	W 14 x 257# x 108" Lg.	HRS
2	1	1" Thk. x 12" Wd. x 18" Lg.	HRS
3	1	1" Thk. x 12" Wd. x 22" Lg.	HRS
4	2	3/4" Thk. x 12-19/32" Wd. x 28-1/2" Lg.	HRS
5	2	3/4" Thk. x 12-19/32" Wd. x 42-1/2" Lg.	HRS

DRILL Ø1-1/32 HOLE THRU
(9) PLACES

DRILL Ø1-1/32 HOLE
THRU (9) PLACES

DRILL Ø25/32 HOLE THRU
(6) PLACES

NOTE:
THIS IS A SAMPLE DRAWING
FOR EDUCATIONAL PURPOSES
ONLY AND IS NOT COMPLETELY
TO THE DEGREE NECESSARY
FOR THE INTENDED PURPOSE
OF THE DISCUSSION OR
REVIEW EXERCISE.

GIRDER BRACE

DATE: 020402   DRAWING NO. W11784S   R1

UNSPECIFIED
FINISH TOLERANCES:
.1X ± 0.020
.1X ± 0.010
.XXX ± 0.005
FRAC. ± 1/32°
ANG. ± 1/2°
SURF. ± 125

REVISION:
1) CHANGED DIM. SCHEME

CHAPTER 14

Practice Exercise 2

Refer to the girder brace in Drawing Number W11784S R1 on page 233 and answer the following questions.

1. How many cutouts are in part number 1?

   _____

2. What type of dimensioning is used to dimension the holes in part number 3?

   _____

3. What type of dimensioning is used to dimension the holes in part number 2?

   _____

4. What is the size of part number 5?

   _____

5. What is the angular cut required on part number 3?

   _____

6. What is the angular cut required on part number 2?

   _____

7. What diameter are the holes in part number 1?

   _____

8. Fully describe the welds required to join part number 1 to part number 4.

   _____

   _____

   _____

9. Is part number 2 welded on both sides?

   _____

10. Is part number 3 welded all the way around?

   _____

11. Are there any tapped holes shown on this drawing?

_____

12. What is the thickness of the web of part number 1?

_____

13. What is the total weight of part number 1?

_____

14. What structural shape is part number 1?

_____

15. What size are the cutouts in the beam?

_____

16. What is the offset from the edge of the flanges to part number 4?

_____

_____

17. What is the distance between part number 2 and part number 3 as they are shown in the right-side view?

_____

18. What is the longest continuous weld required on this assembly?

_____

19. What is the vertical distance between the holes in part number 1?

_____

20. What distance is specified as the horizontal distance between the holes in part number 1?

_____

21. What is the distance between part number 4 and part number 5?

_____

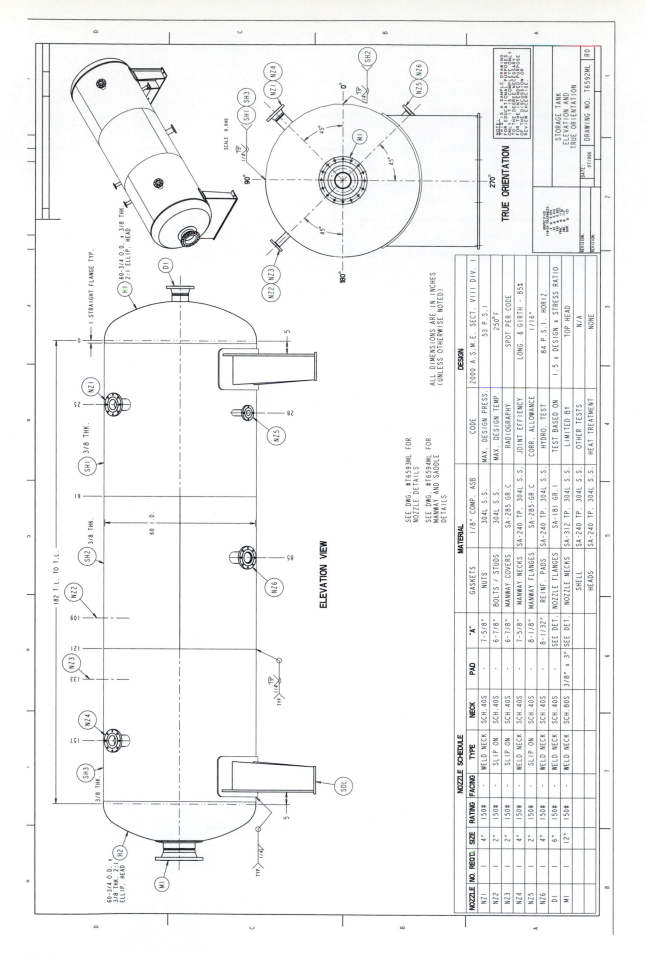

ELEVATION VIEW

TRUE ORIENTATION

SCALE 0.040

SEE DWG. #T6593ML FOR NOZZLE DETAILS

SEE DWG. #T6594ML FOR MANWAY AND SADDLE DETAILS

ALL DIMENSIONS ARE IN INCHES (UNLESS OTHERWISE NOTED)

H1 60-3/4 O.D. x 3/8 THK. 2:1 ELLIP. HEAD

1 STRAIGHT FLANGE TYP.

182 T.L. TO T.L.

60 I.D.

3/8 THK.

60-3/4 O.D. x 3/8 THK. 2:1 ELLIP. HEAD

NOZZLE SCHEDULE

NOZZLE NO.	REQ'D.	SIZE	RATING	FACING	TYPE	NECK	PAD	"A"
NZ1	1	4"	150#	-	WELD NECK	SCH.40S	-	7-5/8"
NZ2	1	2"	150#	-	SLIP ON	SCH.40S	-	6-7/8"
NZ3	1	2"	150#	-	SLIP ON	SCH.40S	-	6-7/8"
NZ4	1	4"	150#	-	WELD NECK	SCH.40S	-	7-5/8"
NZ5	1	2"	150#	-	SLIP ON	SCH.40S	-	8-1/8"
NZ6	1	4"	150#	-	WELD NECK	SCH.40S	-	8-1/32"
D1	1	6"	150#	-	WELD NECK	SCH.40S	-	SEE DET.
M1	1	12"	150#	-	WELD NECK	SCH.80S	3/8" x 3"	SEE DET.

MATERIAL

GASKETS	1/8" COMP. ASB	
NUTS	304L S.S.	
BOLTS / STUDS	304L S.S.	
MANWAY COVERS	SA-285 GR.C	
MANWAY NECKS	SA-240 TP. 304L S.S.	
MANWAY FLANGES	SA-285-GR.C	
REINF. PADS	SA-240 TP. 304L S.S.	
NOZZLE FLANGES	SA-181 GR.1	
NOZZLE NECKS	SA-312 TP. 304L S.S.	
SHELL	SA-240 TP. 304L S.S.	
HEADS	SA-240 TP. 304L S.S.	

DESIGN

CODE	2000 A.S.M.E. SECT. VIII DIV. 1
MAX. DESIGN PRESS.	53 P.S.I.
MAX. DESIGN TEMP.	250°F
RADIOGRAPHY	SPOT PER CODE
JOINT EFFIENCY	LONG. & GIRTH - 85%
CORR. ALLOWANCE	1/16"
HYDRO. TEST	84 P.S.I. HORIZ.
TEST BASED ON	1.5 x DESIGN x STRESS RATIO
LIMITED BY	TOP HEAD
OTHER TESTS	N/A
HEAT TREATMENT	NONE

NOTE: THIS IS A SAMPLE DRAWING FOR DISCUSSION PURPOSES ONLY. IT IS COMPLETE ONLY FOR THE INTENTION FOR THE DESIGN PURPOSE REVIEW EXERCISES OR REVIEW DISCUSSION OR

STORAGE TANK ELEVATION AND TRUE ORIENTATION

DRAWING NO. T6592ML

DATE: 071906

R0

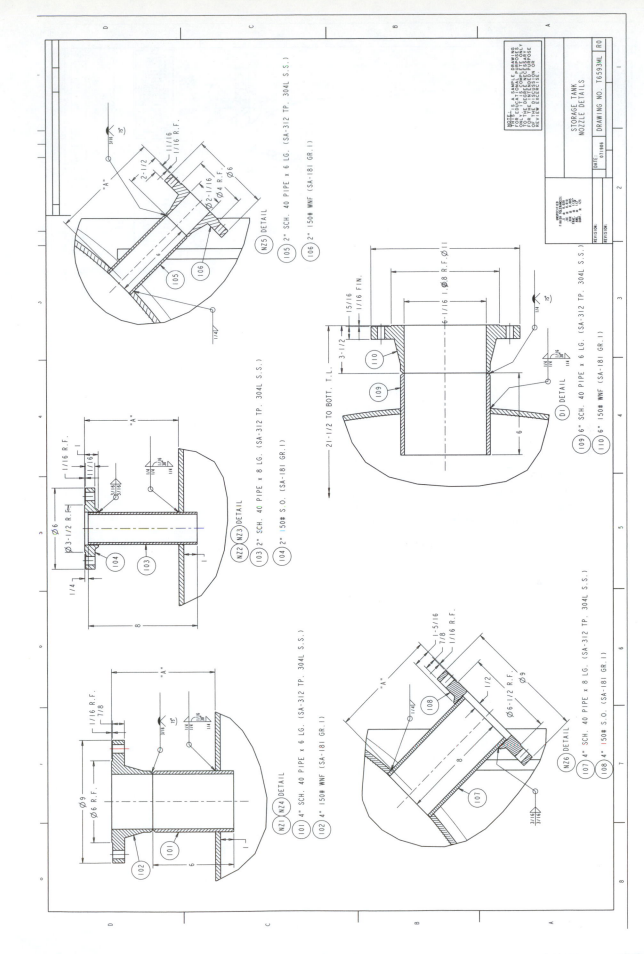

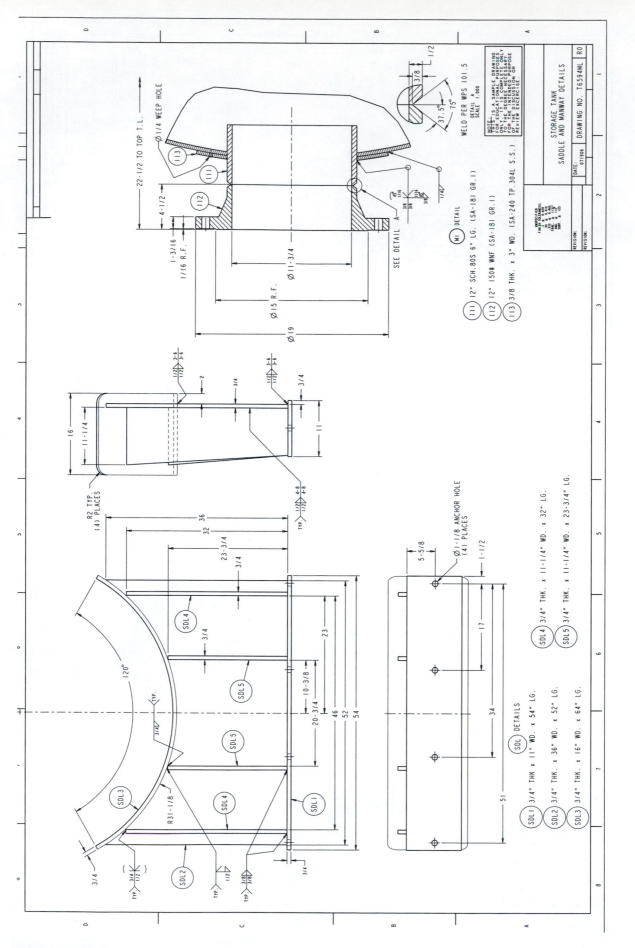

CHAPTER 14
Practice Exercise 3

Refer to Drawing Numbers T6592ML R0 on page 236, T6593ML R0 on page 237, and T6594ML R0 on page 238. Then answer the following questions.

1. What is the joint efficiency of the long seams? _____

2. What is part H2? _____

3. What is the outside diameter (OD) of the tank?_____

4. Specifically describe how part 101 is attached to part 102. _____

   _____

5. Specifically describe how NZ4 is attached to the tank. _____

   _____

   _____

6. What is the distance from the end face of part number 110 to the bottom tangent line (BOTT.T.L.)?

   _____

7. Specifically describe how is SDL2 is attached to SDL1._____

   _____

   _____

8. Specifically describe how SDL5 is attached to SDL3. _____

   _____

   _____

9. What welding procedure specification is called out on this set of drawings? _____

10. What is the width of the straight flange? _____

11. What type of nozzle is NZ5? _____

12. What is the ID of part number 107? _____

Practice Exercise 3 (Continued)

13. What is the outside diameter of the raised face (RF) of part number 104? _____

14. How far does part number 101 extend into the tank? _____

15. How long is part number 107? _____

16. What is the distance from the RF of part number 108 to the end of part number 107? _____

17. What is the purpose of detail A? _____

_____

_____

18. What is the limitation of the tank design? _____

19. How many elliptical heads are there? _____

20. How many rolled sheets are required to make the shell of the tank? _____

21. What type of material is part number 104 made from? _____

22. What is part number 109? _____

23. What is the diameter of the anchor holes in the base of the saddle (SDL)? _____

24. What is the distance between SDL4 and SDL5? _____

25. Specifically describe how SDL3 is attached to SDL2. _____

_____

_____

_____

26. What is the true orientation of NZ6? _____

27. What code does this drawing design refer to? _____

Practice Exercise 3 (Continued)

28. Part number 113 is a reinforcing pad. Specifically describe how it is attached to H2 and how it is attached to the 128" Sched. 80 S Pipe? _____

_____

_____

_____

29. What are the requirements of the hydrotest? _____

30. What is the A dimension of NZ2? _____

31. What is the A dimension of NZ6? _____

32. What radius is called out for the top corners of SDL3? _____

33. Draw the welding symbol used to describe the welds at the circumferential weld seams where the 3/8" thick shell sections are welded together.

34. What is the included angle on the weld joint that joins pieces 109 and 110? _____

35. What size is the land on the weld joint that joins the nozzle neck to the weld neck flange, as shown in detail A of M1 detail? _____

36. What material are bolts and studs to be made from? _____

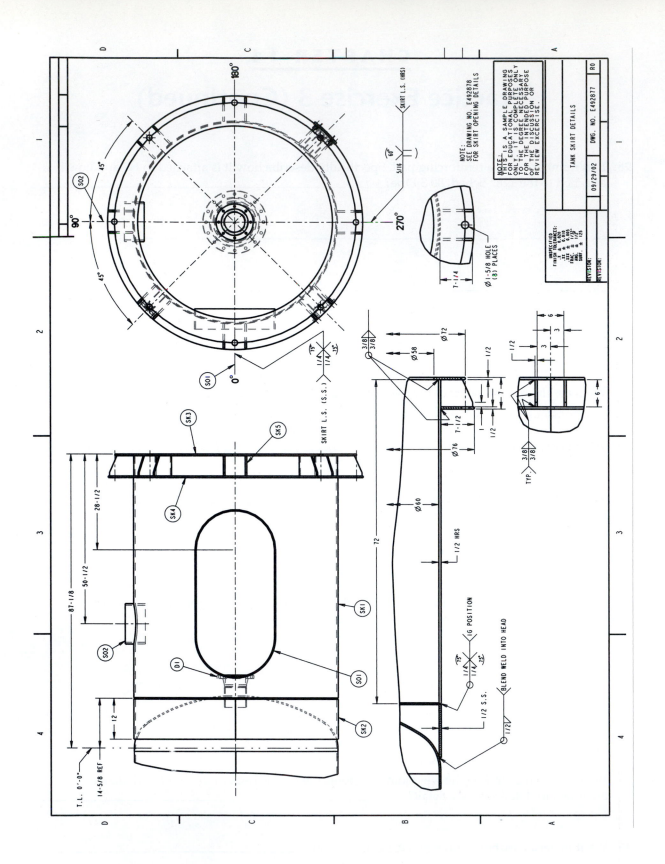

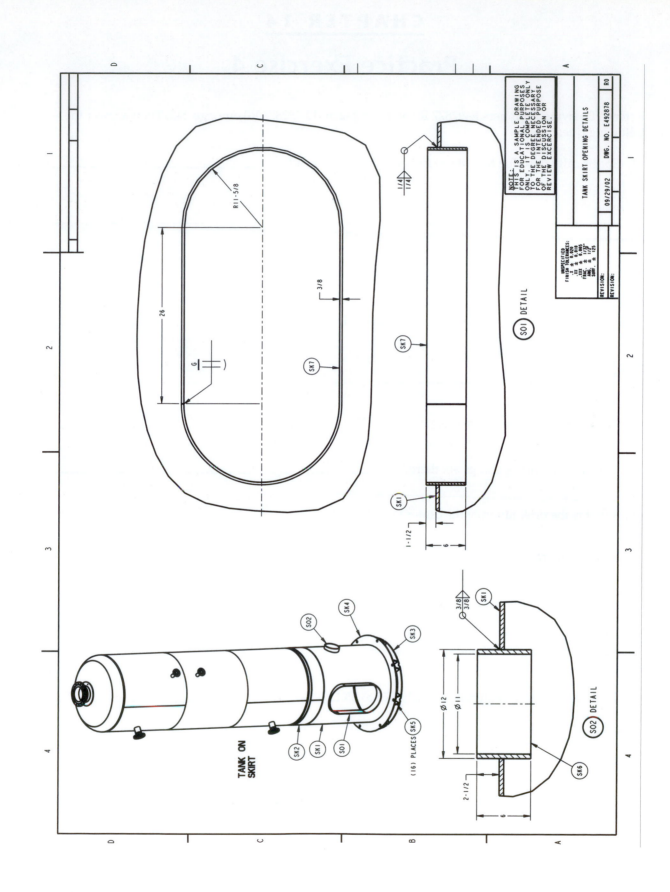

R11-5/8

26

3/8

G

SK7

SK7

SK1

1-1/2

6

SO1 DETAIL

NOTE: THIS IS A SAMPLE DRAWING FOR EDUCATIONAL PURPOSES ONLY. IT IS COMPLETE ONLY FOR THE DEGREE NECESSARY FOR THE DISCUSSION PURPOSE OF THE REVIEW EXERCISE. OR REVIEW EXERCISE.

UNSPECIFIED
FINISH TOLERANCES:
.X ± 0.030
.XX ± 0.010
.XXX ± 0.005
FRAC. ± 1/32
ANG. ± 1/2°
SURF. ± 125

TANK SKIRT OPENING DETAILS

09/29/02   DWG. NO. E492878   R0

REVISION:
REVISION:

SO2

SK4

SK3

1/4
1/4

TANK ON
SKIRT

SK2 SK1 SO1

(16) PLACES SK5

3/8
3/8

SK1

Ø12
Ø11

2-1/2

6

SK6

SO2 DETAIL

CHAPTER 14

Practice Exercise 4

Refer to Drawing Numbers E492877 R0 on page 242 and E482878 R0 on page 243. Then answer the following questions.

1. Specifically describe how the skirt is welded to the head. _____

   _____

   _____

2. What are the thickness, width, and length of SK5? _____

   T _____ W _____ L

3. How many SK5 pieces are there? _____

4. How many skirt openings are there? _____

5. What is the orientation of the center line of SO1? _____

6. What is the ID of SK6? _____

7. How far does SK7 protrude into the inside of the skirt? _____

8. What is the length and width of the SO1 opening? _____

9. Describe the weld required to attach SK5 to SK1. _____

   _____

10. How many total weld joints are covered by the welding symbol located at coordinate A2?

    _____

Practical Exercise 4 (Continued)

11. What is the total length of the skirt? _____

12. How far is T.L. 0'0" from the base of the skirt? _____

13. Are there any holes in SK3? _____

14. What material is SK1 made from? _____

15. Describe in detail, the weld required to join the ends of SK7. _____

_____

_____

16. What is the maximum allowable opening in SK1 to accommodate SK6?

_____

_____

17. What material is SK2 made from? _____

Practical Exercise 4 (Continued)

18. What is located at the bottom center of the head? _____

19. What is the ID of SK4? _____

20. What pieces are to be joined by a weld that is to be made in the 1G position? _____

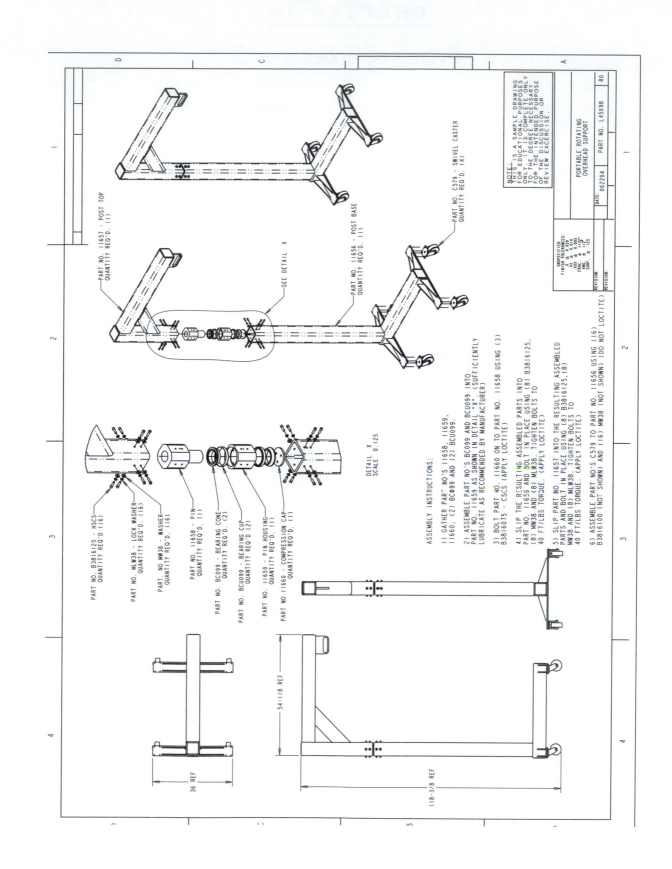

DETAIL X
SCALE 0.125

ASSEMBLY INSTRUCTIONS:

1) GATHER PART NO'S 11658, 11659, 11660, (2) BC099 AND (2) BCU099.

2) ASSEMBLE PART NO'S BC099 AND BCU099 INTO PART NO. 11659 AS SHOWN IN DETAIL "X". (SUFFICIENTLY LUBRICATE AS RECOMMENDED BY MANUFACTURER)

3) BOLT PART NO. 11660 ON TO PART NO. 11658 USING (3) B3816075 - CSCS (APPLY LOCTITE)

4) SLIP THE RESULTING ASSEMBLED PARTS INTO PART NO. 11656 AND BOLT IN PLACE USING (8) B3816125, (8) MW38 AND (8) MLW38. TIGHTEN BOLTS TO 40 FT/LBS TORQUE. (APPLY LOCTITE)

5) SLIP PART NO. 11657 INTO THE RESULTING ASSEMBLED PARTS AND BOLT IN PLACE USING (8) B3816125, (8) MW38 AND (8) MLW38. TIGHTEN BOLTS TO 40 FT/LBS TORQUE. (APPLY LOCTITE)

6) ASSEMBLE PART NO'S C579 TO PART NO. 11656 USING (16) B3816100 (NOT SHOWN) AND (16) MW38 (NOT SHOWN) (DO NOT LOCTITE)

PART NO. B3816125 - HSCS
QUANTITY REQ'D. (16)

PART NO. MLW38 - LOCK WASHER
QUANTITY REQ'D. (16)

PART NO. MW38 - WASHER
QUANTITY REQ'D. (16)

PART NO. 11658 - PIN
QUANTITY REQ'D. (1)

PART NO. BC099 - BEARING CONE
QUANTITY REQ'D. (2)

PART NO. BCU099 - BEARING CUP
QUANTITY REQ'D. (2)

PART NO. 11659 - PIN HOUSING
QUANTITY REQ'D. (1)

PART NO. 11660 - COMPRESSION CAP
QUANTITY REQ'D. (1)

PART NO. 11657 - POST TOP
QUANTITY REQ'D. (1)

PART NO. 11656 - POST BASE
QUANTITY REQ'D. (1)

SEE DETAIL X

PART NO. C579 - SWIVEL CASTER
QUANTITY REQ'D. (4)

54-1/8 REF

36 REF

118-3/8 REF

PORTABLE ROTATING
OVERHEAD SUPPORT

DATE: 062204   PART NO. L45X98   R0

UNSPECIFIED
FINISH TOLERANCES:
.X ± 0.020
.XX ± 0.010
.XXX ± 0.005
FRAC. ± 1/32
ANG. ± 1/2°
SURF. ± .125

REVISION:

REVISION

CHAPTER 14

Practice Exercise 5

Refer to the portable rotating overhead support in Drawing Number L45X98 R0 on page 247. Then answer the following questions.

1. Name the different types of views used on this drawing. _____

   _____

2. Is this a detail or an assembly drawing? _____

3. How many total parts are required to assemble one portable rotating overhead support?

   _____

4. What is the approximate height of part number L45X98? _____

5. How should the bearings be lubricated? _____

6. What is part number 11659? _____

7. How are the swivel casters to be attached to the post base? Be specific. _____

   _____

8. Which part numbers require Locktite? _____

   _____

9. What is the part number for the portable rotating overhead support? _____

10. According to the assembly instructions, how many parts must be assembled prior to the installation of part number 11657 (post top)? _____

   _____

CHAPTER 14

Math Supplement

Comprehensive Student Review

Add the following. Write your answer in lowest form.

1. $\dfrac{3}{8} + \dfrac{3}{8} =$

2. $\dfrac{7}{16} + \dfrac{5}{16} =$

3. $\dfrac{2}{64} + \dfrac{7}{16} =$

4. $1\dfrac{1}{4} + 5\dfrac{1}{4} =$

5. $7\dfrac{23}{32} + \dfrac{9}{16} =$

6. $5\dfrac{15}{32} + 3\dfrac{3}{8} =$

Change the following fractions from improper fractions to mixed numbers. Show the fractions in their lowest form.

7. $\dfrac{7}{4} =$

8. $\dfrac{13}{2} =$

9. $\dfrac{10}{8} =$

10. $\dfrac{82}{64} =$

11. $\dfrac{44}{32} =$

12. $\dfrac{81}{64} =$

Use subtraction to answer the following questions. Write your answers in lowest form.

13. $\dfrac{3}{4} - \dfrac{1}{4} =$

14. $\dfrac{9}{16} - \dfrac{5}{16} =$

15. $\dfrac{47}{64} - \dfrac{17}{64} =$

16. $8\dfrac{5}{8} - 5\dfrac{1}{8} =$

17. $3\dfrac{27}{32} - \dfrac{9}{16} =$

18. $10\dfrac{1}{8} - 2\dfrac{7}{32} =$

Use a calculator to convert the following fractions from common fraction form to decimal fraction form.

19. $\dfrac{3}{4} =$

20. $\dfrac{9}{16} =$

21. $\dfrac{3}{32} =$

22. $\dfrac{5}{8} =$

23. $\dfrac{27}{32} =$

24. $\dfrac{1}{8} =$

25. $\dfrac{11}{16} =$

26. $\dfrac{3}{8} =$

27. $\dfrac{13}{16} =$

Write the following measurements in their lowest form.

```
        ┌──────────────── 28 ────────────────┐
        ┌─── 29 ───┐
16ths                                    1F
         1        11          12         13        14
32nds
        └─ 30 ─┘
        └──────────── 31 ────────────┘
        └──────────── 32 ──────────┘
```

28. _____

29. _____

30. _____

31. _____

32. _____

33. A line dimensioned at $27\frac{1}{2}''$ is shown on a drawing with a scale of $\frac{3}{8} = 1$.
    What is the length of the line?

    _____

34. A line dimensioned at $2\frac{3}{16}''$ is shown on a drawing with a scale of 3:1.
    What is the length of the line?

    _____

35. A line dimensioned at $6'11''$ is shown on a drawing with a $\frac{1}{4}$ scale.
    What is the length of the line?

    _____

36. What is the diameter of a circle that has a radius of 15 inches? _____

37. What is the radius of a circle that has a diameter of 17.4 inches? _____

38. What is the diameter of the following circle? _____

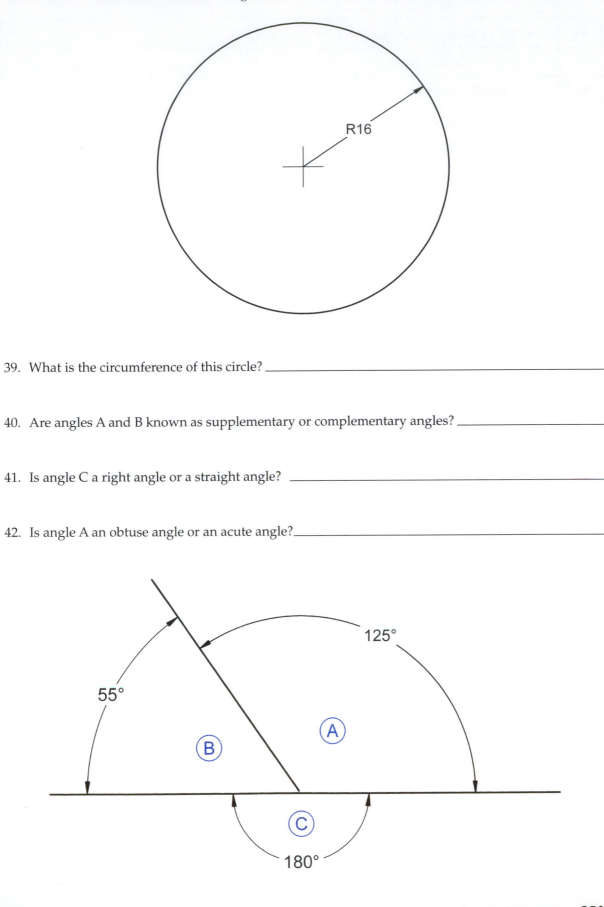

R16

39. What is the circumference of this circle? _____

40. Are angles A and B known as supplementary or complementary angles? _____

41. Is angle C a right angle or a straight angle? _____

42. Is angle A an obtuse angle or an acute angle? _____

125°

55°

A

B

C

180°

Convert the following.

43.	61°30′ to decimal degrees	
44.	118°15′ to decimal degrees	
45.	38.7 mm to inches	
46.	22″ to millimeters	
46.	7.8 meters to inches	
48.	0.8125″ to cm	
49.	4000 mm to meters	
50.	$46\frac{3}{8}''$ to mm	

Acronyms, Abbreviations, and Letter Designations

- This appendix contains a list of typical acronyms, abbreviations, and letter designations that may be used on drawings and specifications. Some are used often on drawings that most welding personnel will work with, and some are used more seldomly. Many of the abbreviations are used for more than one term. The person interpreting the drawing determines the meaning of the acronym by understanding the context in which it is being used.

- Note that this list does not include all acronyms, abbreviations, and letter designations that may be used on drawings. It also does not ensure an abbreviation or acronym that is used stands for any of the terms listed below.

Table A.1

A	ampere, area	AL	aluminum
AAW	air acetylene welding	ALIGN	alignment
AB	anchor bolt	ALLOW	allowance
AB	adhesive bonding	ALT	alternate
ABRSV	abrasive	ALY	alloy
ABT	about	ANL	anneal
ABW	arc braze welding	ANOD	anodize
AC	arc cutting	ANSI	American National Standards Institute
ACCESS	accessory		
ACCUMR	accumulator	APPD	approved
ACET	acetylene	APPROX	approximately
ACR	across	API	American Petroleum Institute
ACT	actual	AR	as Required
ACTR	actuator	ASME	American Society of Mechanical Engineers
AD	adaptive control		
ADD	addendum	ASP	arc spraying
ADH	adhesive	ASSEM	assemble
ADPT	adapter	ASSY	assembly
ADJ	adjust, adjustable	ASTM	American Society for Testing and Materials
AET	acoustic emission test		
AHW	atomic hydrogen welding	AU	automatic
AISI	American Iron & Steel Institute	AUTH	authorized

(continued)

253

AUTO	automatic	CAC-A	air carbon arc cutting
AUX	auxiliary	CAD	computer aided drafting
AVG	average	CALC	calculated
AW	arc welding	CAM	computer aided manufacturing
AWG	American Wire Gauge	CANC	cancelled
AWS	American Welding Society	CAP	capacity
		CAP SCR	cap screw
B	brazing, butt	CARB	carburize
B&S	brown & sharpe	CAW	carbon arc welding
BB	back to back	CAW-G	gas carbon arc welding
BB	block brazing	CAW-S	shielded carbon arc welding
B/B	back to back	CAW-T	twin carbon arc welding
BBE	bevel both ends	CBORE	counterbore
BCD	bolt circle diameter	C/C	center to center
BMAW	bare metal arc welding	CCW	counter clockwise
BE	both ends	CC	color code
BET	between	CDRILL	counterdrill
BEV	bevel	CDS	cold drawn steel
BHN	brinell hardness number	CENT	centrifugal
BL	baseline, bendline	CEW	co-extrusion welding
BLD	blind	CFM	cubic feet per minute
BLK	black, blank, block	CFS	cold finished steel, cubic feet per second
BLT	bolt		
BM	beam	CG	center of gravity
BM	bench mark	CH	case harden
B/M	bill of material	CHAM	chamfer
BNH	burnish	CHAN	channel
BOE	bevel one end	CHG	change
BOT	bottom	CHK	check
BP	base plate	CI	cast iron
B/P	blueprint	CIM	computer integrated manufacturing
BR	bend radius		
BRG	bearing	CIR	circle, circular
BRK	break	CIRC	circumference
BRKT	bracket	CJP	complete joint penetration
BRO	broach	CKT	circuit
BRS	brass	CL	centerline, clearance
BRZ	bronze	CLP	clamp
BRZG	brazing	CM	centimeter
BS	british standards	CMM	coordinate measuring machine
BSC	basic		
Btm	bottom	CNC	computer numerical control
BUSH	bushing	CO	company
BW	braze welding	COL	column
BWG	birmingham wire gauge	COMB	combination
		CONC	concentric
C	Celsius, centigrade, channel, Chipping, corner	COND	condition
		CONN	connect, connection, connector
C to C	center to center		
CA	corrosion allowance	CONT	control
CABW	carbon arc braze welding	COP	copper
CAC	carbon arc cutting	COV	cover

CPLG	coupling	E	edge, edge Joint
C/R	chamfer or radius	E to E	edge to edge, end to end
CRS	cold rolled steel	EA	each
CR VAN	chrome vanadium	EBBW	electron beam braze welding
CS	carbon steel, cast steel	EBC	electron beam cutting
CSA	canadian standards association	EBW	electron beam welding
C'SINK	countersink	EBW-HV	high vacuum electron beam welding
CSK	countersink		
CSTG	casting	EBW-MV	medium vacuum electron beam welding
CTD	coated		
CTR	center, contour	EBW-NV	non-Vacuum electron beam welding
Cu	copper, cubic		
CuFt	cubic feet	ECC	eccentric
CV	check valve, constant voltage	ECO	engineering change order
CW	clockwise, cold welding	EFF	effective
CWB	canadian welding bureau	EGW	electrogas welding
CYL	cylinder, cylindrical	EL	elevation
		ELL	elbow
D	diameter	ELEC	electric
DAT	datum	ELEV	elevation
DB	dip brazing	ENCL	enclosure
DBC	diameter bolt circle	ENG	engine, engineer
DBL	double	ENGR	engineer, engineering
DCN	document change notice	ENGRG	engineering
DCN	drawing change notice	EO	engineering order
DEC	decimal	EQ	equal
DECR	decrease	EQL SP	equally spaced
DEG	degree	EQUIV	equivalent
DET	detail	EST	estimate
DEV	develop, developed, deviation	ESW	electroslag welding
DF	drop forge	ESW-CG	consumable guide electroslag welding
DFB	diffusion brazing		
DFT	draft	ET	electromagnetic test
DFW	diffusion welding	EX	extra
DIA	diameter	EXB	exothermic brazing
DIAG	diagonal, diagram	EXBW	exothermic braze welding
DIM	dimension	EXP	expansion, experimental
DIN	German Institute for Standardization	EXT	extension, external
		EXW	explosion welding
DISC	disconnect		
DIST	distance	F	Fahrenheit
DL	developed length	FAB	fabricate
DN	diameter number, down	FAO	finish all over
DO	ditto	F&D	flanged & dished
DP	deep, depth	FB	furnace brazing
DR	drill, drill rod	FCAW	flux cored arc welding
DRM	drafting room manual	FCAW-G	gas-shielded flux cored arc welding
DS	dip soldering		
DSGN	design	FCAW-S	self-shielded flux cored arc welding
DUP	duplicate		
DVTL	dovetail	FDRY	foundry
DW	developed width	FF	flat face
DWG	drawing	FHD	flat head
DWN	drawn		

(*continued*)

FIG	figure	H	hammering
FIL	fillet, fillister	H&G	harden and grind
FIM	full indicator movement	HCS	high carbon steel
FIN	finish	HD	head
FIX	fixture	HDL	handle
FL	fluid	HDLS	headless
FLB	flow brazing	HDN	harden
FLEX	flexible	HDW	hardware
FLG	flange	HEBC	high energy beam cutting
FLOW	flow welding	HEX	hexagon
FLSP	flame spraying	HGR	hanger
FLSP-W	wire flame spraying	HGT	height
FORG	forge, forging	HH	hex head
FOW	forge welding	HIPW	hot isostatic pressure welding
F/P	flat pattern	HLS	holes
FR	frame, front	HOR	horizontal
FRW	friction welding	HPW	hot pressure welding
FRW-DD	direct drive friction welding	HR	hot rolled, hour
		HRA	Rockwell hardness A
FRW-I	inertia friction welding	HRB	Rockwell hardness B
FS	far side	HRC	Rockwell hardness C
FS	furnace soldering	HRD	Rockwell hardness D
FST	forged steel	HRE	Rockwell hardness E
FSW	friction stir welding	HRF	Rockwell hardness F
FTG	fitting, footing	HRG	Rockwell hardness G
FURN	furnish	HRS	hot rolled steel
FW	fillet weld, flash welding	HS	high speed
FWD	forward	HSG	housing
FXD	fixed	HSS	high speed steel, hollow structural steel
G	grind	HT	heat treat
GA	gage, gauge	HT TR	heat treat
GAL	gallon	HV	vickers hardness
GALV	galvanized	Hv	vickers hardness
GDR	girder	HVOF	high velocity oxy-fuel spraying
GEN	generator		
GIRD	girder	HVY	heavy
GMAC	gas metal arc cutting	HYD	hydraulic
GMAW	gas metal arc welding	HYDRO	hydrostatic
GMAW-P	gas metal arc welding—pulsed	IB	induction brazing
GMAW-S	gas metal arc welding—short circuit transfer	ID	inside diameter
		IDENT	identification
GOSL	gauge outstanding leg	ILLUS	illustration
GR	grade	IN	inch, inches
GRD	grind, ground	INCL	inclined, include, included, including, inclusive
GRV	groove		
GSKT	gasket	INCR	increase
GTAC	gas tungsten arc cutting	IND	indicator
GTAW	gas tungsten arc welding	INDEP	independent
GTAW-P	gas tungsten arc welding—pulsed	INFO	information
		INS	iron soldering

INSP	inspect, inspection		LTR	letter
INST	instruct, instrument		LUB	lubricate
INSTL	install, installation		LWN	long weld neck
INT	interior, internal, intersect			
IR	inside radius		M	machine, magnaflux, metric
IRB	infrared brazing		MA	manual
IRS	infrared soldering		MACH	machine
IS	induction soldering		MAG	magnesium
ISO	International Organization for Standardization		MAINT	maintenance
			MAJ	major
IW	induction welding		MALL	malleable
			MAN	manual
JCT	junction		MAT	material
JNL	journal		MAT'L	material
JOG	joggle		MAX	maximum
JT	joint		MB	machine bolt
			MC	miscellaneous channel
K	Kelvin, key		ME	mechanized
kg	kilogram		MECH	mechanical, mechanism
KNRL	knurl		MED	medium
KP	kick plate		MFG	manufacturing
KST	keyseat		Mg	magnesium
KWY	keyway		MI	malleable iron
			MIAW	magnetically impelled arc welding
L	lap (lap joint)		MID	middle
LAB	laboratory		MIL	military
LAQ	lacquer		MIN	minimum, minute
LAM	laminate		MISC	miscellaneous
LAT	lateral		MK	mark
LB	pound		ML	mold line
LBC	laser beam cutting		mm	millimeter
LBC-A	laser beam air cutting		MMC	maximum material condition
LBC-EV	laser beam evaporative cutting		MOD	modification
LBC-IG	laser beam inert gas cutting		MOT	motor
LBC-O	laser beam oxygen cutting		MS	machine steel, mild steel
LBL	label		MT	magnetic particle test
LBS	pounds		MTG	mounting
LBW	laser beam welding		MTL	material
LC	low carbon		MULT	multiple
LG	length, long		MW	manway
LH	left hand			
LIN	linear		N	newton
LIQ	liquid		N/A	not applicable
LL	lower limit		NC	national coarse, numerical control
L/M	list of material		NEC	national electrical code
LMC	least material condition		NEG	negative
LOA	length overall		NF	national fine
LOC	locate		NL	nosing line
LR	long radius		NO	number
LS	limit switch		NOM	nominal, nomenclature
LT	leak test, light			
LTD	limited			

(*continued*)

NORM	normalize		PL	parting line, places, plate
NPS	nominal pipe size		PLSTC	plastic
NPSM	national pipe thread for		PLT	pilot, plate
	straight mechanical Joints		PNEU	pneumatic
NPT	national pipe thread		PNL	panel
NRT	neutron radiographic test		POE	plain one end
NS	near side, nickel steel		POL	polish
NTS	not to scale		POS	position, positive
			PR	pair
OA	overall		PRESS	pressure
OAC	oxygen arc cutting		PRI	primary
OAW	oxyacetylene welding		PROC	process, procedure
OBS	obsolete		PROD	product, production
OC	on center, oxygen cutting		PROJ	project, projection
O/C	on center		PRT	proof test
OC-F	flux cutting		PSI	pounds per square inch
OC-P	metal powder cutting		PSIG	pounds per square inch gage
OD	outside diameter		PSP	plasma spraying
OFC	oxyfuel gas cutting		PT	part, penetrant test, point
OFC-A	oxyacetylene cutting		PV	plan view
OFC-H	oxyhydrogen gas cutting		PW	projection welding
OFC-P	oxypropane cutting			
OFW	oxyfuel welding		QA	quality assurance
OG	oxygen gouging		QC	quality control
OHW	oxyhydrogen welding		QAL	quality
OLC	oxygen lance cutting		QTR	quarter
O/O	out to out		QTY	quantity
OPP	opposite		QUAL	quality
OPPH	opposite hand			
OPTL	optional		R	radius, remove, rolling
OR	outside radius		RA	Rockwell hardness A
ORIG	original		RAD	radius
OSL	outstanding leg		RB	resistance brazing, rockwell
OZ	ounce			hardness B
			RC	Rockwell hardness C
P	pitch, port		RD	round
PAC	plasma arc cutting		RD	running dimension
PAT	pattern		RECP	receptacle
PATT	pattern		RECT	rectangle
PAW	plasma arc welding		REF	reference
PBE	plain both ends		REG	regular, regulator
PC	piece, pitch circle		REINF	reinforce
PCH	punch		REL	release, relief
PCS	pieces		REM	remove
PD	pitch diameter		REPAD	reinforcing pad
PERF	perforate, perforated		REQ'D	required
PERM	permanent		RET	retainer, return
PERP	perpendicular		REV	reverse, revision, revolution
PEW	percussion welding		RF	raised face
PF	press fit		RFS	regardless of feature size
PFD	preferred		RGH	rough
PGW	pressure gas welding		RH	right hand, rockwell hardness
PKG	package, packaging		RIV	rivet

RM	ream	SPG	spring
RND	round	SPHER	spherical
RO	robotic	SPL	splice, special
ROW	roll welding	SPRKT	sprocket
RP	reference point	SQ	square
RS	resistance soldering	SR	short radius
RSEW	resistance seam welding	SS	stainless steel
RSEW-HF	high frequency seam welding	SS	shop stock
RSEW-I	induction seam welding	S/S	seam to seam
RSEW-MS	mash seam welding	SST	stainless steel
RSW	resistance spot welding	SSW	solid state welding
RT	radiographic test	STD	standard
RW	resistance welding	STK	stock
RW-PC	pressure controlled resistance welding	STL	steel
		STR	straddle, straight, strip
		STR	straight
S	schedule, second, soldering, standard (Beam)	SUB	substitute
		SUP	supply, support
SA	semiautomatic	SURF	surface
SAE	society of automotive engineers	SW	arc stud welding
		SYM	symbol, symmetrical
SAW	submerged arc welding	SYS	system
SAW-S	submerged arc welding— series arc	T	tee joint, teeth, tooth, TEE (Beam), top
SCH	schedule	T&B	top and bottom
SCR	screw	TAB	tabulate
SE	semi-elliptical	TAN	tangent
SEC	second, section	TB	torch brazing
SECT	section	TBE	threaded both ends
SEP	separate	T&C	threaded and coupled
SEQ	sequence	TC	thermal cutting
SER	serial, series	TCAB	twin carbon arc brazing
SERR	serrate	TECH	technical
SF	slip fit, spot face, straight flange	TEMP	template, temporary
		THD	thread
SFT	shaft	THK	thick
SGL	single	THSP	thermal spraying
SH	sheet	TIR	total indicator reading
SI	International System of Units	TOC	top of concrete, table of contents
SILS	silver solder		
SL	slide	TOE	threaded one end
SLMS	seamless	TOL	tolerance
SLV	sleeve	TOR	torque
SM	seam	TOS	top of steel
SMAC	shielded metal arc cutting	TOT	total
SMAW	shielded metal arc welding	TPF	taper per foot
SO	slip on	TPI	threads per inch, taper per inch
SOC	socket		
SO GR	soft grind	TPR	taper
SOL	solenoid	TS	tensile strength, tool steel, torch soldering,
SP	space, spaces, spare		
SPA	space, spaces, spacing	TS	tubesheet
SPEC	specification	TYP	typical
SPECS	specifications		

(continued)

U	unspecified	VOL	volume
UL	upper limit	VS	verses
U/N	unless noted	VT	visual test
UNC	unified national coarse		
UNEF	unified national extra fine	W	watt, width, wide, wideflange, tungsten
UNF	unified national fine		
UNIV	universal	WASH	washer
UOS	unless otherwise specified	WD	working drawing
USS	ultrasonic soldering	WDF	woodruff
USW	ultrasonic welding	WI	wrought iron
UT	ultrasonic test	WL	working line
UW	upset welding	WLD	weld, welded
UW-HF	high frequency upset welding	WN	weld neck
		W/O	without
UW-I	induction upset welding	WP	wear plate, weatherproof, working point
V	volt	WRT	wrought
VAC	vacuum	WT	weight
VAR	variable		
VER	vernier	YP	yield point
VERT	vertical	YS	yield strength
VD	void		
VIB	vibrate	Z	ZEE (structural shape)

Common Fraction, Decimal Fraction, and Millimeter Conversions

Common Fraction of an Inch	Decimal Fraction of an Inch	Millimeters	Common Fraction of an Inch	Decimal Fraction of an Inch	Millimeters
1/64	0.0156250	0.3968750	33/64	0.5156250	13.0968750
1/32	0.0312500	0.7937500	17/32	0.5312500	13.4937500
3/64	0.0468750	1.1906250	35/64	0.5468750	13.8906250
1/16	0.0625000	1.5875000	9/16	0.5625000	14.2875000
5/64	0.0781250	1.9843750	37/64	0.5781250	14.6843750
3/32	0.0937500	2.3812500	19/32	0.5937500	15.0812500
7/64	0.1093750	2.7781250	39/64	0.6093750	15.4781250
1/8	0.1250000	3.1750000	5/8	0.6250000	15.8750000
9/64	0.1406250	3.5718750	41/64	0.6406250	16.2718750
5/32	0.1562500	3.9687500	21/32	0.6562500	16.6687500
11/64	0.1718750	4.3656250	43/64	0.6718750	17.0656250
3/16	0.1875000	4.7625000	11/16	0.6875000	17.4625000
13/64	0.2031250	5.1593750	45/64	0.7031250	17.8593750
7/32	0.2187500	5.5562500	23/32	0.7187500	18.2562500
15/64	0.2343750	5.9531250	47/64	0.7343750	18.6531250
1/4	0.2500000	6.3500000	3/4	0.7500000	19.0500000
17/64	0.2656250	6.7468750	49/64	0.7656250	19.4468750
9/32	0.2812500	7.1437500	25/32	0.7812500	19.8437500
19/64	0.2968750	7.5406250	51/64	0.7968750	20.2406250
5/16	0.3125000	7.9375000	13/16	0.8125000	20.6375000
21/64	0.3281250	8.3343750	53/64	0.8281250	21.0343750
11/32	0.3437500	8.7312500	27/32	0.8437500	21.4312500
23/64	0.3593750	9.1281250	55/64	0.8593750	21.8281250
3/8	0.3750000	9.5250000	7/8	0.8750000	22.2250000
25/64	0.3906250	9.9218750	57/64	0.8906250	22.6218750
13/32	0.4062500	10.3187500	29/32	0.9062500	23.0187500
27/64	0.4218750	10.7156250	59/64	0.9218750	23.4156250
7/16	0.4375000	11.1125000	15/16	0.9375000	23.8125000
29/64	0.4531250	11.5093750	61/64	0.9531250	24.2093750
15/32	0.4687500	11.9062500	31/32	0.9687500	24.6062500
31/64	0.4843750	12.3031250	63/64	0.9843750	25.0031250
1/2	0.5000000	12.7000000	1	1.0000000	25.4000000

Table B-1

APPENDIX C

Millimeter, Decimal Fraction, and Common Fraction Conversions

Millimeters	Decimal Fraction of an Inch	Common Fraction of an Inch	Millimeters	Decimal Fraction of an Inch	Common Fraction of an Inch
0.10	0.0039370		10.00	0.3937008	
0.20	0.0078740		11.00	0.4330709	
0.30	0.0118110		11.11	0.4375000	7/16
0.397	0.0156250	1/64	12.00	0.4724409	
0.40	0.0157480		12.70	0.5000000	1/2
0.50	0.0196850		13.00	0.5118110	
0.60	0.0236220		14.00	0.5511811	
0.70	0.0275591		14.29	0.5625000	9/16
0.79	0.0312500	1/32	15.00	0.5905512	
0.80	0.0314961		15.88	0.6250000	5/8
0.90	0.0354331		16.00	0.6299213	
1.00	0.0393701		17.00	0.6692913	
1.59	0.0625000	1/16	17.46	0.6875000	11/16
2.00	0.0787402		18.00	0.7086614	
3.00	0.1181102		19.00	0.7480315	
3.18	0.1250000	1/8	19.05	0.7500000	3/4
4.00	0.1574803		20.00	0.7874016	
4.76	0.1875000	3/16	20.64	0.8125000	13/16
5.00	0.1968504		21.00	0.8267717	
6.00	0.2362205		22.00	0.8661417	
6.35	0.2500000	1/4	22.23	0.8750000	7/8
7.00	0.2755906		23.00	0.9055118	
7.94	0.3125000	5/16	23.81	0.9375000	15/16
8.00	0.3149606		24.00	0.9448819	
9.00	0.3543307		25.00	0.9842520	
9.53	0.3750000	3/8	25.40	1.0000000	1

Fraction and Decimal Equivalents

Sixteenths	1/16	2/16	3/16	4/16	5/16	6/16	7/16	8/16	9/16	10/16	11/16	12/16	13/16	14/16	15/16	16/16
Eights		1/8		2/8		3/8		4/8		5/8		6/8		7/8		8/8
Quarters				1/4				2/4				3/4				4/4
Halves								1/2								2/2
Wholes																1
Decimal Equivalent	.062	.125	.375	.250	.312	.375	.438	.500	.562	.625	.687	.750	.812	.875	.938	1.00

Table D-1

Welded and Seamless Wrought Steel Pipe Size Chart for Selected Pipe Sizes within the NPS range of 1/8″ through 20″*

NPS†	O.D.	Wall Thickness, Where Applicable, for Schedule 5 Through 160, STD, XS and XXS													
		5	10	20	30	40	60	80	100	120	140	160	STD	XS	XXS
$\frac{1}{8}$	0.405	—	.049	—	.057	.068	—	.095	—	—	—	—	.068	.095	—
$\frac{1}{4}$	0.540	—	.065	—	.073	.088	—	.119	—	—	—	—	.088	.119	—
$\frac{3}{8}$	0.675	—	.065	—	.073	.091	—	.126	—	—	—	—	.091	.126	—
$\frac{1}{2}$	0.840	.065	.083	—	.095	.109	—	.147	—	—	—	.188	.109	.147	.294
$\frac{3}{4}$	1.050	.065	.083	—	.095	.113	—	.154	—	—	—	.219	.113	.154	.308
1	1.315	.065	.109	—	.114	.133	—	.179	—	—	—	.250	.133	.179	.358
$1\frac{1}{4}$	1.660	.065	.109	—	.117	.140	—	.191	—	—	—	.250	.140	.191	.382
$1\frac{1}{2}$	1.900	.065	.109	—	.125	.145	—	.200	—	—	—	.281	.145	.200	.400
2	2.375	.065	.109	—	.125	.154	—	.218	—	—	—	.344	.154	.218	.436
$2\frac{1}{2}$	2.875	.083	.120	—	.188	.203	—	.276	—	—	—	.375	.203	.276	.552
3	3.500	.083	.120	—	.188	.216	—	.300	—	—	—	.438	.216	.300	.600
$3\frac{1}{2}$	4.000	.083	.120	—	.188	.226	—	.318	—	—	—	—	.226	.318	—
4	4.500	.083	.120	—	.188	.237	—	.337	—	.438	—	.531	.237	.337	.674
5	5.563	.109	.134	—	—	.258	—	.375	—	.500	—	.625	.258	.375	.750

*This chart includes information found in ASME B36.10M-2004. It is for educational purposes only and should not be used for design or other industrial requirements. It is not to be construed as a complete listing of pipe sizes for welded and seamless wrought steel pipe.

Wall Thickness, Where Applicable, for Schedule 5 Through 160, STD, XS and XXS

NPS[†]	O.D.	5	10	20	30	40	60	80	100	120	140	160	STD	XS	XXS
6	6.625	.109	.134	—	—	.280	—	.432	—	.562	—	.719	.280	.432	.864
8	8.625	.109	.148	.250	.277	.322	.406	.500	.594	.719	.812	.906	.322	.500	.875
10	10.750	.134	.165	.250	.307	.365	.500	.594	.719	.844	1.000	1.125	.365	.500	1.000
12	12.750	.156	.180	.250	.330	.406	.562	.688	.844	1.000	1.125	1.312	.375	.500	1.000
14	14.000	.156	.250	.312	.375	.438	.594	.750	.938	1.094	1.250	1.406	.375	.500	—
16	16.000	.165	.250	.312	.375	.500	.656	.844	1.031	1.219	1.438	1.594	.375	.500	—
18	18.000	.165	.250	.312	.438	.562	.750	.938	1.156	1.375	1.562	1.781	.375	.500	—
20	20.000	.188	.250	.375	.500	.594	.812	1.031	1.281	1.500	1.750	1.969	.375	.500	—

[†]Key: NPS = nominal pipe size

O.D. = outside diameter

STD = standard

XS = extra strong

XXS = double extra strong

— = not applicable

ASME Y14.5 and ISO
Symbol Comparison Chart

SYMBOL FOR:	ASME Y14.5	ISO
DIMENSION ORIGIN	⌀→	⌀→
FEATURE CONTROL FRAME	⊕ Ø0.5Ⓜ A B C	⊕ Ø0.5Ⓜ A B C
CONICAL TAPER	▷	▷
SLOPE	◁	◁
COUNTERBORE	⌴	NONE
SPOTFACE	⌴SF⌴	NONE
COUNTERSINK	⌵	NONE
DEPTH/DEEP	⤋	NONE
SQUARE	□	□
DIMENSION NOT TO SCALE	15	15
NUMBER OF PLACES	8X	8X
ARC LENGTH	⌒105	⌒105
RADIUS	R	R
SPHERICAL RADIUS	SR	SR
SPHERICAL DIAMETER	SØ	SØ
CONTROLLED RADIUS	CR	NONE
BETWEEN	* ↔	↔ (proposed)
STATISTICAL TOLERANCE	⟨ST⟩	NONE
CONTINUOUS FEATURE	⟨CF⟩	NONE
DATUM TARGET	Ø6/A1 or ⊙Ø6/A1	Ø6/A1 or ⊙Ø6/A1
MOVABLE DATUM TARGET	⟀A1	⟀A1 (proposed)
TARGET POINT	✕	✕
* May be filled or not filled		

SYMBOL FOR:	ASME Y14.5	ISO
STRAIGHTNESS	—	—
FLATNESS	⟋⟍	⟋⟍
CIRCULARITY	○	○
CYLINDRICITY	⌀/	⌀/
PROFILE OF A LINE	⌒	⌒
PROFILE OF A SURFACE	⌓	⌓
ALL AROUND	⟋○	⟋○
ALL OVER	⟋◎	⟋◎ (proposed)
ANGULARITY	∠	∠
PERPENDICULARITY	⊥	⊥
PARALLELISM	//	//
POSITION	⊕	⊕
CONCENTRICITY (Concentricity and Coaxiality in ISO)	◎	◎
SYMMETRY	≡	≡
CIRCULAR RUNOUT	*↗	*↗
TOTAL RUNOUT	*↗↗	*↗↗
AT MAXIMUM MATERIAL CONDITION	Ⓜ	Ⓜ
AT MAXIMUM MATERIAL BOUNDARY	Ⓜ	NONE
AT LEAST MATERIAL CONDITION	Ⓛ	Ⓛ
AT LEAST MATERIAL BOUNDARY	Ⓛ	NONE
PROJECTED TOLERANCE ZONE	Ⓟ	Ⓟ
TANGENT PLANE	Ⓣ	NONE
FREE STATE	Ⓕ	Ⓕ
UNEQUALLY DISPOSED PROFILE	Ⓤ	UZ (proposed)
TRANSLATION	▷	NONE
DIAMETER	⌀	⌀
BASIC DIMENSION (Theoretically Exact Dimension in ISO)	50	50
REFERENCE DIMENSION (Auxiliary Dimension in ISO)	(50)	(50)
DATUM FEATURE	*⟙A⟘	*⟙ or *⟙A⟘

* May be filled or not filled

APPENDIX

G

Master Chart of Welding and Joining Processes

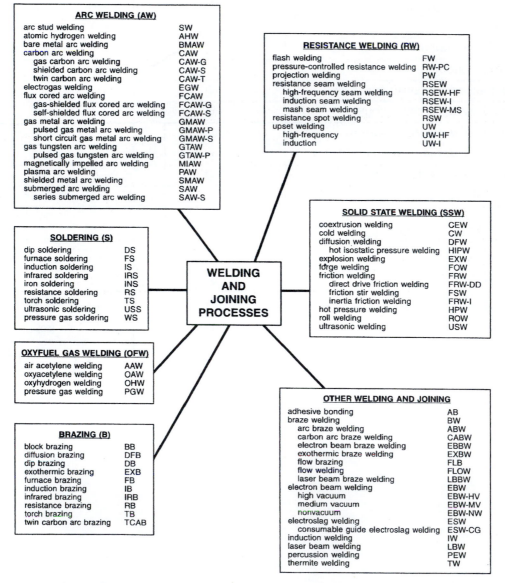

ARC WELDING (AW)

arc stud welding	SW
atomic hydrogen welding	AHW
bare metal arc welding	BMAW
carbon arc welding	CAW
gas carbon arc welding	CAW-G
shielded carbon arc welding	CAW-S
twin carbon arc welding	CAW-T
electrogas welding	EGW
flux cored arc welding	FCAW
gas-shielded flux cored arc welding	FCAW-G
self-shielded flux cored arc welding	FCAW-S
gas metal arc welding	GMAW
pulsed gas metal arc welding	GMAW-P
short circuit gas metal arc welding	GMAW-S
gas tungsten arc welding	GTAW
pulsed gas tungsten arc welding	GTAW-P
magnetically impelled arc welding	MIAW
plasma arc welding	PAW
shielded metal arc welding	SMAW
submerged arc welding	SAW
series submerged arc welding	SAW-S

RESISTANCE WELDING (RW)

flash welding	FW
pressure-controlled resistance welding	RW-PC
projection welding	PW
resistance seam welding	RSEW
high-frequency seam welding	RSEW-HF
induction seam welding	RSEW-I
mash seam welding	RSEW-MS
resistance spot welding	RSW
upset welding	UW
high-frequency	UW-HF
induction	UW-I

SOLDERING (S)

dip soldering	DS
furnace soldering	FS
induction soldering	IS
infrared soldering	IRS
iron soldering	INS
resistance soldering	RS
torch soldering	TS
ultrasonic soldering	USS
pressure gas soldering	WS

SOLID STATE WELDING (SSW)

coextrusion welding	CEW
cold welding	CW
diffusion welding	DFW
hot isostatic pressure welding	HIPW
explosion welding	EXW
forge welding	FOW
friction welding	FRW
direct drive friction welding	FRW-DD
friction stir welding	FSW
inertia friction welding	FRW-I
hot pressure welding	HPW
roll welding	ROW
ultrasonic welding	USW

WELDING AND JOINING PROCESSES

OXYFUEL GAS WELDING (OFW)

air acetylene welding	AAW
oxyacetylene welding	OAW
oxyhydrogen welding	OHW
pressure gas welding	PGW

BRAZING (B)

block brazing	BB
diffusion brazing	DFB
dip brazing	DB
exothermic brazing	EXB
furnace brazing	FB
induction brazing	IB
infrared brazing	IRB
resistance brazing	RB
torch brazing	TB
twin carbon arc brazing	TCAB

OTHER WELDING AND JOINING

adhesive bonding	AB
braze welding	BW
arc braze welding	ABW
carbon arc braze welding	CABW
electron beam braze welding	EBBW
exothermic braze welding	EXBW
flow brazing	FLB
flow welding	FLOW
laser beam braze welding	LBBW
electron beam welding	EBW
high vacuum	EBW-HV
medium vacuum	EBW-MV
nonvacuum	EBW-NW
electroslag welding	ESW
consumable guide electroslag welding	ESW-CG
induction welding	IW
laser beam welding	LBW
percussion welding	PEW
thermite welding	TW

Figure 54A—Master Chart of Welding and Joining Processes

Source: AWS A3.0:2001, Figure 54A, reproduced with permission from the American Welding Society (AWS), Miami, FL.

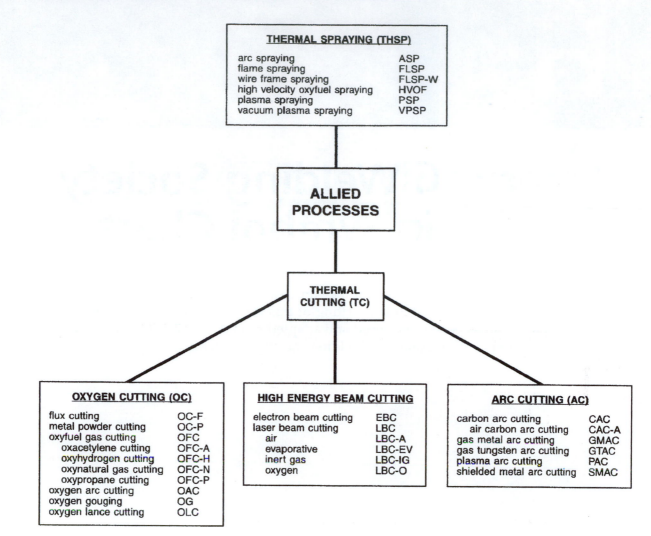

Figure 54B—Master Chart of Allied Processes

Source: AWS A3.0:2001, Figure 54B, reproduced with permission from the American Welding Society (AWS), Miami, FL.

American Welding Society Welding Symbol Chart

Appendix H is the version of the American Welding Society Welding Symbol Chart that is current at the writing of this text. It includes the following two pages. It has been reproduced and adapted from AWS A2.4:2007 Annex E with permission of the American Welding Society (AWS), Miami, FL.

American Welding Society Welding Symbol Chart

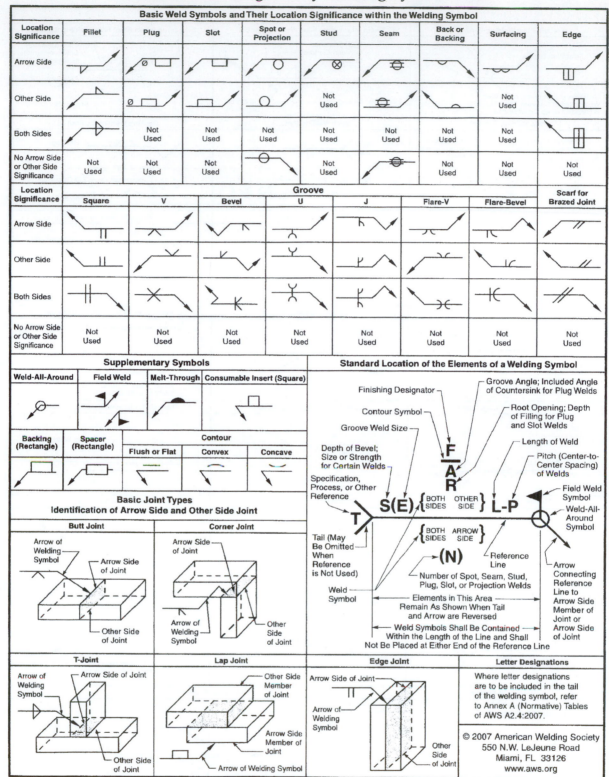

Source: AWS A2.4:2007, Annex E reproduced and adapted with permission from the American Welding Society (AWS), Miami, FL.

American Welding Society Welding Symbol Chart

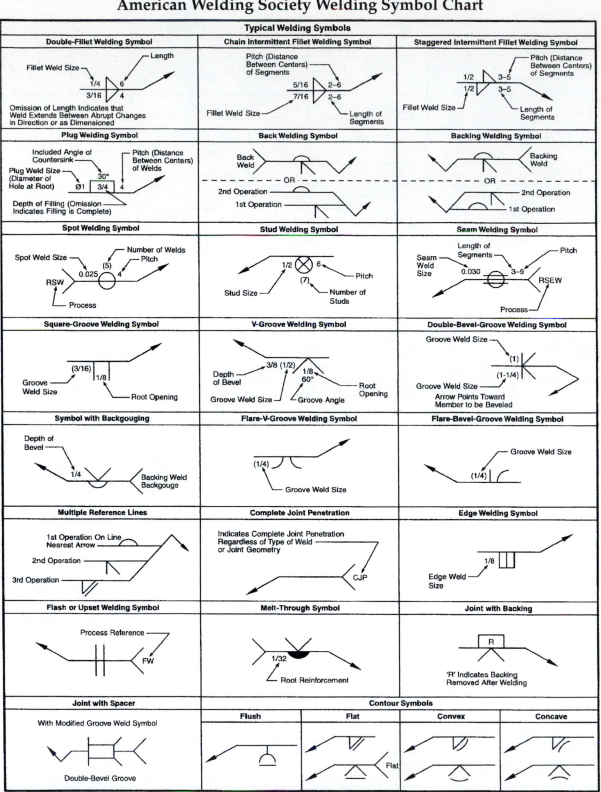

Source: AWS A2.4:2007, Annex E reproduced and adapted with permission from the American Welding Society (AWS), Miami, FL.

arrow side: The side of the weld joint that the arrow is pointing to.

assembly drawing: A drawing that shows how parts are assembled.

auxiliary view: A profile view of a slanted surface shown at 90 degrees to the standard orthographic projection view of the surface.

backing weld: A weld that is made on the back side of a weld joint to provide backside reinforcement (backing) prior to the weld made on the front side of the weld joint.

back weld: A weld made on the back side of a weld joint after the weld has been made on the front side.

balloon: A closed shape that contains text for the purpose of identifying individual parts on assemblies.

baseline dimensioning: A method of dimensioning where dimensions originate from a single base line and not from the ends of other dimensions.

bevel: The angled edge of a plate or hole.

bevel angle: The angle of a bevel.

bill of material: A list of materials needed. Also known as *materials list*.

blind hole: A hole that doesn't go all the way through the part.

blue-line print: A particular type of copy of an engineering drawing where the lines of the design are shown in blue.

blueprint: The name of a specific type of copy of a drawing made with a process that produces a blue background with drawing lines shown in white.

bolt circle: A circle on which a number of bolts or holes have their center.

bore: Machined enlargement of a hole.

boss: A raised circular area (circular pad) to provide added strength or other benefits.

burr: The sharp edge left on a part after cutting, machining, drilling, or other method of material removal.

CAD: An acronym for computer-aided drafting.

callout: An annotation that consists of text with a leader; typical annotations include information on hole sizes, thread specifications, chamfers, etc.

center line: Thin line with alternating short and long dashes used to locate the center of holes, arcs, and other symmetrical objects.

chain dimensioning: *See* conventional dimensioning.

chain line: A thick line with alternately spaced short and long dashes used to indicate the location and extent of an area.

chamfer: A beveled edge connecting two surfaces.

clearance: The amount of distance between two mating parts.

clearance hole: A hole that is slightly larger than the item required to be put through it.

concentric: Having a common center or common axis.

conventional dimensioning: A method of dimensioning where all dimensions in any one direction for an object do not originate from a single extension line. Also known as *chain dimensioning*.

coordinate system: A system whereby numbers and letters are placed around the perimeter of a drawing sheet for the purpose of providing a location grid.

counterbore: Enlargement of the end of a hole; typically used so that the outer head of a bolt can be flush or recessed from the surface of the part.

countersink: Angled outer edge on a hole so that a flathead-type fastener fits flush into the hole properly.

cutting plane line: A line used to indicate where the view is to be theoretically cut for the section view.

datum: A reference assumed to be theoretically exact from where measurements are made; the reference may be a point, an axis, or a plane.

deburr: To remove the sharp edge left on a part after cutting, machining, drilling, or other method of material removal.

degree: An angular unit of measurement for $\frac{1}{360}$ of a circle.

destructive testing: Any testing method that requires destruction of the part or a portion of the part.

detail drawing: A drawing that shows all the detail, dimensions, etc., to make a part.

dimension: A measurement used to show size or location.

dimension line: A line located between extension lines, centerlines, other lines, or combinations thereof to show the extent of a dimension.

drawing sheet: A standard size sheet with a border and title block; used to provide a drawing space for an engineering design or part of an engineering design.

elevation: A view that shows height.

extension line: A fine thin line extending from but not touching the object in the view to give location points for dimensioning.

face: *v.*, to make a surface flat or smooth; *n.*, the front surface of a weld.

field weld: A weld that is made somewhere other than the original fabrication area.[1] Also referred to as a *site weld*.

fillet: A concave inside radius.

fillet weld: A triangular shaped weld placed in a weld joint between two perpendicular surfaces.

finish mark: A symbol placed on a surface located on a drawing to indicate that the surface is to be finished in some manner.

gage: An instrument used for measurement. Variant spelling of the word gauge.

gauge: A number assigned to sheet metal to describe its thickness or to wire size to describe its diameter. Variant spelling of the word gage.

general note: A note placed on a drawing that gives information that is to be applied to the whole drawing or whole set of drawings.

geometric dimensioning and tolerancing (GD&T): A very precise method of dimensioning and tolerancing that uses a set of symbols for characteristics such as flatness, straightness, circularity, angularity, runout, etc.

heat treat: The use of heat to transform specific properties in metals.

---

[1] American Welding Society (AWS) A2 Committee on Definitions and Symbols, AWS A3.0:2001 Standard Terms and Definitions, 14.

hidden line: A line made of short and long dashes alternately spaced, and used to show the edges of an object hidden in the view drawn.

included angle: The total angle of a grooved weld joint.

inseparable assembly drawing: An assembly drawing where the individual parts are to be permanently joined together.

isometric drawing: A type of pictorial drawing that is made at 30-degree angles from a point located on a horizontal line.

leader: A line with an arrow, dot, or slash at the end that points to or attaches to a place or surface of an object on the drawing.

least material condition (LMC): The condition where the least amount of material exists for a part within its acceptable tolerance zone.

limit: The minimum or maximum a dimension can be after adding the applicable tolerance to the basic dimension.

linear dimension: The dimension for a straight line.

local note: Note placed on a drawing that applies to a specific, localized area or component of the drawing. A local note overrides a general note for the specific case in point.

long break line: Type of line used to indicate a theoretical removal of part of a drawing.

main assembly drawing: The overall assembly drawing for the complete design.

materials list: *See* bill of material.

maximum material condition (MMC): The condition where the maximum amount of material exists for a part within its acceptable tolerance zone.

melt-through: "Visible root reinforcement produced in a joint welded from one side."[2]

mono-detail drawing: A detail drawing that consists of only one part.

multi-detail drawing: A drawing that consist of more than one individually detailed part.

nominal size: The named size of an object; may be slightly different from the actual size.

   Example 1: 6" schedule 40 pipe actually measures 6.065" I.D. and 6.625" O.D.
   Example 2: A 2 × 4 wood stud actually measures $1^{1}/2$" × $3^{1}/2$".

non-destructive examination (NDE): Any inspection or testing method that does not destroy the object being examined.

non-destructive testing (NDT): Any testing method that does not destroy the material being tested.

note: information on a drawing typically given under the heading "NOTE"; *See also* local note; general note; specification.

object line: *See* visible line.

oblique drawing: A type of pictorial drawing made with the front view in line with a horizontal line and either the left or right side typically oriented at a 30 or 45-degree angle to the front view; the width may be shown at shorter than its true width.

orthographic projection drawing: A drawing consisting of views located at 90 degrees to each other that show each side of an object from a straight-on viewpoint.

overlay: *See* surfacing.

pad: A raised area typically used to provide additional material where bolt holes are required; a nonstandard term used for surfacing or overlay.

---

[2] AWS A3.0:2001 Standard Welding Terms and Definitions, 25.

phantom line: Thin line made up of a series of one long and two short dashes, and used for the purpose of showing alternate positions, mating or adjacent parts, or repeated detail.

pictorial drawing: A drawing that gives a three-dimensional representation from the viewpoint of the observer; pictorial drawings are typically either oblique or isometric.

pitch: The center-to-center spacing between intermittent welds; the distance between corresponding points on two gear teeth on a gear; the spacing of threads on a bolt, etc.

plan view: Another name for the top view in an orthographic projection drawing.

plotter: Machine for making printouts of drawings.

print: Copy of a drawing.

radius: The distance of a straight line beginning at the center of a circle or sphere and ending at any point on the outside surface of the circle or sphere. The radius is half the diameter of a circle (D/2).

reference dimension: A dimension that can be derived from other dimensions given on the drawing but is given to aid in reading the drawing. Reference dimensions should be shown within parentheses and should not be used for making accurate measurements during construction, manufacturing, or inspection.

S-break line: Non-standard term used in this text to describe conventional representation of breaks used for cylindrical objects such as pipe and bar stock.

scale: The ratio between the distances on the drawing and the distances in reality.

section lines: Thin lines drawn at an angle used to indicate the cut area in the section view of an object.

section view: A view that shows the object as if it were cut and opened up.

short break line: Irregular wavy line drawn to indicate a short cutout area of a view.

site weld: *See* field weld.

six principal views: Front, top, right side, left side, bottom, and back views in an orthographic projection drawing.

specification: A note that gives precise information on material, welding process, filler material, electrode, etc.

spotface: A machined flat surface around the outside surface of a hole for the purpose of providing a flat area for a bolt, nut, or washer seating.

stud weld: Type of weld where a stud (screw, bolt, etc.) is welded to a surface.

surfacing: The application of material to a surface for the purpose of building up the surface to a desired dimension or surface property.

symmetrical: Having two halves that are the same. Symmetric parts typically have a center line that indicates the line of symmetry.

tabular dimensioning: A method of dimensioning where dimensions are listed within a table.

tangent line: A straight line that contacts a circle or curve at only one point.

tap: *v.* the act of making internal threads in an object; *n.* tool used to make internal threads in an object.

taper: A gradual decrease in the thickness or diameter of an object.

title block: The ruled area of a drawing that gives the name of the drawing, drawing number, date drawn, scale, etc.

tolerance: Amount of variance from a dimension that is acceptable.

top drawing: A term used for the first drawing in a drawing package; typically shows the overall assembly of the design. May also be called the main assembly drawing.

typical: Normally abbreviated typ. for the purpose of signifying that the instructions given are to be the same for any other places on the drawing that match the current location.

viewing plane line: A type of line with arrows used to indicate a viewpoint along a specific plane within a particular view.

visible line: Line showing an edge of an object that is visible to the eye in the particular view drawn. Also known as an *object line*.

welding procedure specification (WPS): The instructions that give the information on how to make a weld.

welding symbol: A symbol used to give information on the location of a weld, the type and size of weld, and other supplementary information that describes the requirements of the weld. See Figure 10-1.

weld joint: The junction where pieces are welded together.

weld symbol: The symbol that is placed on the reference line of the welding symbol that indicates the type of weld to be made.

whiteprint: The name of a specific type of copy of a drawing made with a process that produces a white background with drawing lines shown in blue or black.

zoning coordinate system: The use of numbers and letters around the perimeter of a drawing so that specific areas of the drawing can be located easily.

INDEX